Particle Acceleration and Detection

The series "Particle Acceleration and Detection" is devoted to monograph texts dealing with all aspects of particle acceleration and detection research and advanced teaching. The scope also includes topics such as beam physics and instrumentation as well as applications. Presentations should strongly emphasize the underlying physical and engineering sciences. Of particular interest are

- contributions which relate fundamental research to new applications beyond the immediate realm of the original field of research
- contributions which connect fundamental research in the aforementioned fields to fundamental research in related physical or engineering sciences
- concise accounts of newly emerging important topics that are embedded in a broader framework in order to provide quick but readable access of very new material to a larger audience.

The books forming this collection will be of importance to graduate students and active researchers alike.

More information about this series at http://www.springer.com/series/5267

Valery D. Shemelin • Sergey A. Belomestnykh

Multipactor in Accelerating Cavities

Valery D. Shemelin
Cornell University
Ithaca, NY, USA

Sergey A. Belomestnykh
Fermi National Accelerator Laboratory
Batavia, IL, USA

ISSN 1611-1052 ISSN 2365-0877 (electronic)
Particle Acceleration and Detection
ISBN 978-3-030-48197-1 ISBN 978-3-030-48198-8 (eBook)
https://doi.org/10.1007/978-3-030-48198-8

This Springer imprint is published by the registered company Springer Nature Switzerland AG
The registered company address is: Gewerbestrasse 11, 6330 Cham, Switzerland

Preface

Multipactor is a discharge phenomenon that commonly occurs in and limits the performance of radio frequency (RF) structures operating under vacuum. Such structures are used in satellites, plasma and fusion experiments, and particle accelerators. The authors have many years of experience developing and operating RF systems for particle accelerators. Our interest in the subject of multipacting is driven by a desire to design and build better particle accelerators, which dates back to our years of study at university. Over the years, we have dealt with multipactor at different stages of accelerator developments: designing RF structures so that they are less susceptible to multipactor, studying the phenomenon experimentally, and developing approaches to mitigate it when encountered in operations.

This book aims to summarize our experience and systematically present results scattered over many journal articles and conference proceedings. To a large extent the book is based on our own experimental, theoretical, and computer simulations body of work. However, it is not limited to our research only, as we strive to present a complete picture of multipactor in accelerating cavities. Works of other authors are discussed when appropriate and, in several cases, new explanations of experimental results are offered. The approaches presented in the book are applicable to both normal conducting and superconducting RF (SRF) structures, although most of the examples relate to SRF cavity structures.

While multipactor in evacuated transmission lines (waveguides, coaxial lines, etc.) and in the presence of dielectrics (e.g., ceramic RF windows separating gas-filled transmission lines from the evacuated part of a structure or dielectric-lined structures) can be a major performance limitation, these areas are not covered in the book. We shall deal only with the origins of multipactor and shall therefore not consider any saturation effects.

The readers should note that the terms "multipactor" and "multipacting" are used interchangeably in this book.

Structure of the Book

The book starts with an introductory overview of the multipactor effect with some historical observations and a brief overview of how our understanding of multipactor has developed over time, followed by considerations of the most common aspects of this phenomenon.

Part I is devoted to examination of multipactor in a planar gap. The simple geometry of two infinite planes separated by a gap is very instrumental in understanding multipactor. One can derive analytical formulae for equations of motion, stability conditions, and other constraints and then find boundaries of the discharge zones, and this is presented in Chap. 2. A pioneer experiment by A. J. Hatch and H. B. Williams is discussed in detail and the deviation between experiments and the theory is explained in Chap. 3, which deals with a generalized phase stability describing more complicated particle dynamics in the gap, which can exist beyond the limits predicted by the simple stability condition explained in the previous chapter. One such more complicated mode of multipactor, the so-called "ping-pong mode", is of particular interest and is considered in Chap. 4. Finally, Chap. 5 briefly introduces computer codes used for simulation of multipactor in a planar gap.

Treatment of multipactor in crossed RF fields of accelerating cavities is presented in Part II. After introduction of accelerating radio frequency cavities in Chap. 6, Chap. 7 considers experimental study of multipactor in a normal conducting 430 MHz cavity and theoretical work on inclusion magnetic field in the equations of motion. Chapter 8 describes an approach to analyze multipacting near the cavity equator – a case of special importance for SRF elliptical cavities. Chapter 9 provides a general analytic approach to treating one-point multipacting in superconducting cavities.

Though computer simulation has become the current most popular tool for predicting multipactor barriers in RF cavities, we show in this book that the analytic approach in some cases explains the phenomena better. Examples of such results are the influence of weld seams (Sects. 8.8 and 8.9) and explanation of the apparently non-resonant but distinctly defined discharge bands in Sect. 9.5.

Different configurations of the cavity geometry and fields create conditions for different kinds of multipactor, for example, one-point or two-point multipactor, which can be of the 1st, 2nd, etc., order. We offer a general approach to the description of the discharge with the help of two parameters: geometric parameter and field parameter. Actually, this description was presented in the very first works on multipactor theory in a flat gap: zones of multipactor in coordinates fd, the product of the frequency and gap size as the geometric parameter, and the amplitude of the voltage U between the plates as the field parameter. We generalize this description to other kinds of multipactor.

Part III provides insights on how one can design multipacting-free cavities (Chap. 10) and transitions between the cavities and beam pipes (Chap. 11). For a cavity, time-consuming simulations of multipactor can be replaced by calculation of fields in the area where multipacting can possibly occur. These field calculations, typically performed with electromagnetic field eigensolvers, are much faster and give us the ability to quickly evaluate different variations of the geometry under consideration. For a transition between cavity and beam pipe, the solution is to avoid any local minimum of the surface electric field.

Acknowledgements

The authors would like to thank people who helped us with preparation of this book at various stages: Hasan Padamsee, without whom this work could not be possible, Dmitri Miakishev and Slava Yakovlev for the SuperLANS program, Ursula van Rienen and Rongli Geng for their support and interest in the work, Mingqi Ge for the samples of weld seams, Niels LaWhite for his help in editing, and Slava Yakovlev for reading the manuscript and providing valuable comments. We would like to thank Editage (www.editage.com) for English language editing. Our special thanks go to Springer's editorial team for their valuable help in the publication process.

Last but not least, we are grateful to our wives Ludmila (V. S.) and Natasha (S. B.) for their support and patience in this endeavor.

Ithaca, NY, USA — Valery D. Shemelin
Batavia, IL, USA — Sergey A. Belomestnykh
March, 2020

The original version of the book has been revised. A correction to this book can be found at https://doi.org/10.1007/978-3-030-48198-8_12

Contents

Chapter 1
Introductory Overview

Multipactor is a phenomenon of resonant multiplication of electrons in a vacuum under the influence of a radio frequency (RF) electromagnetic field, which determines the particles' trajectories. Upon impacting the surface, "primary" electrons create secondary particles. Depending on the energy of primaries and the surface material, the ratio of secondary to primary electrons can be greater than unity. This ratio is called secondary electron yield (SEY). If the time of flight of electrons between impacts is in resonance with oscillations of the RF field, an electron avalanche can be initiated.

The multipactor effect was first observed by French physicist Camille Gutton in 1924 [1] and has been studied in subsequent years by him and his son as a discharge in ionized gas [2, 3]. However, this phenomenon was identified as a secondary emission discharge and further studied in 1934 by Philo T. Farnsworth [4], the inventor of electronic television, who attempted to take advantage of the effect in an amplifier. He coined the word "multipactor" when he first invented a vacuum tube under that name in 1934 [5]. Later the use of the term multipactor was extended and became associated with the effect itself. A multipactor (or "AC Electron Multiplier") first was a device where the electrons make *multiple impacts* on metal electrode surfaces, freeing the secondary electrons. If there is a high-frequency electric field in the volume of the device, the electrons can absorb energy from this field under certain conditions and their number can increase like an avalanche due to secondary emission. It was one of few attempts "to tame" this phenomenon; however, also see an ion source described by H. Alfvén and H.-J. Cohn-Peters in [6] and the multipactor electron gun by W. J. Gallagher [7], with the original conception due to Maury Tigner. These inventions did not find wide use. Moreover, nowadays the multipactor effect has become an obstacle to be avoided in normal operation of accelerating structures and RF windows, vacuum electronics, radars, satellite communication devices, etc.

As stated in [6], later investigation by N. E. Backmark and U. Bengston led to a theoretical analysis by U. Danielsson in which he proposed a resonance breakdown

V. D. Shemelin, S. A. Belomestnykh, *Multipactor in Accelerating Cavities*, Particle Acceleration and Detection, https://doi.org/10.1007/978-3-030-48198-8_1

mechanism for the motion of the secondary electrons. In his model, the electrons cross the space between two electrodes in an odd number of RF half-periods. Prediction of the multipacting region (or zone) boundaries using this model was in a rough agreement with experiments.

We can summarize earlier investigations and distinguish two mechanisms that govern whether multipactor can be sustained. The first mechanism is due to the dynamics of electrons inside an RF device: the particle trajectories should be in resonance with the RF field and stable with respect to perturbations. In other words, there should exist "attractor" trajectories. Any perturbed trajectories – be they due to fluctuation of the impact time (RF phase), distribution of the initial energies, or the emission angles of the secondary electrons – should eventually evolve toward an attractor trajectory.

The second mechanism is related to the physical properties of surface materials, which determine secondary emission. The SEY is a function of the kinetic energy of primary electrons, the incident angle, the type of material, and the condition of its surface. Figure 2.2 shows a typical velocity distribution of the secondary electrons. In general, three types of secondary electrons are distinguished. First, a certain portion of the primary electrons is back-scattered elastically. These particles are represented by a narrow maximum near the velocity of the primary electrons. Then, some are scattered from several atoms inside the material before being reflected back out. These are called "rediffused" electrons and are represented by an area between the two maxima in Fig. 2.2. Finally, the remaining particles interact with the material in a more complicated way and create "true secondary electrons". The position of the broad high maximum of true secondary electrons is nearly independent of the energy of the primary electrons [8].

Typically, for a given frequency and geometry of an RF device, multipactor can be sustained in several ranges of applied RF voltage, which are often called multipacting zones or bands. Boundaries of the multipacting zones can be determined by the secondary emission yield, even in presence of resonance conditions, or by a stability condition when a deviation of RF phase leads to further deviation and finally results in electrons falling out of resonance with the oscillating RF field and thus to a low impact energy with SEY less than unity.

The initial velocity of the secondary electrons has been recognized as an important factor for the correct calculation of multipactor. As R. Kishek et al. stated in their historical review [9], "Gill and von Engel (Ref. [10] in this book) introduced the ad hoc assumption that a parameter k, equal to the impact velocity of the primaries relative to the emission velocity of the secondaries, is constant. There is no physical basis for this assumption. Furthermore, Gill and von Engel's attempt at using the theory to interpret their experimental results was flawed, and it was not until Hatch and Williams reformulated the theory in the 1950s to explain their own multipactor experiments (Ref. [11, 12] in this book) that reasonable agreement was obtained. Since Hatch and Williams retained the constant "k" assumption of Gill and von Engel, the modified theory became known as "the constant-k" theory and for decades remained the classic theory on the accessibility of multipactor ... More recently, in 1988, Vaughan (Ref. [13] in this book) has promoted an alternative to the constant-k theory ... derived from first principles. Vaughan's theory replaced the baseless assumption of constant k with the more realistic assumption of a monoenergetic nonzero initial velocity."

In fact, the nonzero initial velocity and the abandoning of the constant-k theory was introduced as early as in 1955 by Krebs and Meerbach [14]. In 1986, this assumption was used by V. Shemelin [15] to find analytically limits of multipactor zones defined by return of electrons to the electrode from which they started. In combination with the condition of stability, upper and lower boundaries of the multipactor zones were found. These results, with some corrections, are presented in Chap. 2.

Vaughan deserves the credit for the first explicit expression for saturated power dissipation in the simple two-surface case. He also described different manifestations of multipactor, offered a short analysis of multipactor in the presence of transverse magnetic field, and drew attention to the increase of SEY in the case of oblique impacts of primary electrons on the surface of an electrode. Some authors have used an empirical formula by Vaughan for the SEY vs. energy of primary electrons [16]. However, contemporary computer codes use an advanced probabilistic electron emission model by Furman and Pivi [17].

Over the years, wide usage of RF devices has revealed many different varieties of multipactor discharge and stimulated their theoretical treatment. Complex hybrid resonance modes and a non-resonant "polyphase regime" have been described [18]. Multipactor with periodically repeating start phases [19], one-surface multipactor in the presence of a DC electric bias [20] or a permanent magnetic field [21], and multipactor in crossed RF electric and magnetic field when both fields are important in spite of a relatively low velocity of electrons [22, 23] are examples of this development. Another interesting case is that of devices operating with two-frequency signals. Mixing of two frequencies can break the conditions of existence for the discharge [24]. There are many interesting cases that have been studied or are yet to be investigated.

Multipactor in a flat gap, as the simplest and most common kind of this phenomenon, will be analyzed in detail because it is easier to show methods of analysis on this simplest geometry. While flat gaps in the cavities are not a common situation, some parts of the accelerating structures, such as couplers, are close to this geometry. The solutions presented here can be applied to other forms of multipactor. After analysing multipactor in a flat gap in Part I, the remainder of the book will be devoted to multipactor in resonant RF cavities.

To offer an insight into the whole picture of the phenomenon here, in the Introduction, we briefly review multipacting in different devices and modes of operation.

Multipactor can be a major limitation to the maximum RF power transmitted to RF cavities via input couplers. The input coupler, or fundamental power coupler, is an interface between a transmission line, carrying RF power from a source at the fundamental frequency, and a cavity [25]. For efficient transmission of RF power, the coupler is designed as an impedance-matching device. In addition, as the transmission line is typically gas-filled and the cavity operates under ultra-high vacuum, the coupler must incorporate an RF-transparent vacuum barrier – a ceramic RF window.

Multipactor can occur on metal surfaces and on the surfaces of dielectrics [9]. In the latter case, it is especially difficult to model multipactor because of the high SEY of most dielectrics and due to charge build-up on their surfaces. Unmitigated multipactor on dielectric RF windows can lead to devastating consequences, such as the window cracking, opening a vacuum leak and contaminating the cavity.

There are various ways to minimize the risk of multipacting in couplers by influencing either the dynamics of electron motion or physical properties of the surface. For example, one can avoid dangerous multipacting zones (i) by choosing the right dimensions and impedance of the transmission line; (ii) by adding geometric features, e.g., grooved surfaces [26], that disrupt the multipacting trajectories and thus effectively reduce SEY; and (iii) by using materials or a coating with low SEY. To reduce the high SEY of ceramics, RF windows are often coated with a thin layer (several nm) of low-SEY material, such as titanium nitride, (e.g., see [27]). Another method of suppressing multipacting in transmission lines is to destabilize the electron motion by applying external DC electric or magnetic fields. These external bias fields can have either a beneficial or detrimental effect depending on the field direction and/or polarity. The DC bias voltage applied between the inner and outer conductors has been used successfully in many coaxial input couplers to mitigate multipactor.

As the input couplers are typically based on either coaxial lines or rectangular waveguides, many efforts have been dedicated to study multipactor in such transmission lines. As the literature in this areas is very extensive, we only mention several relatively recent and impactful references here.

In the pioneering paper [28], the authors investigated the dynamics of electron trajectories in coaxial lines using 2D computer code. Their results showed that the multipacting power in standing wave (SW) operation quite accurately obeys the following scaling laws for one-point and two-point multipacting:

$$P_{one\ point} \sim (fD)^4 Z, \quad P_{two\ point} \sim (fD)^4 Z^2, \tag{1.1}$$

where f is the RF frequency, D is the diameter of the outer conductor, and Z is the line impedance. A "susceptibility" plot was generated based on the simulations. Also, it was shown that in traveling wave (TW) operation, the multipacting levels shift according to the simple rule $P_{TW} = 4P_{SW}$. Further analytical, numerical, and experimental investigations followed, with some results reported in Refs. [29, 30].

Important investigations of rectangular waveguides have been performed at Cornell University in the framework of developing methods to suppress multipactor in input couplers for superconducting cavities [31–34]. In particular, two multipactor suppression methods were proposed and demonstrated experimentally [33]: the slotted waveguide method and DC magnetic bias method. The Cornell investigations were followed by more theoretical and numerical studies and numerical simulations, e.g., [35–37].

Beam-induced multiplication of electrons resulting in so-called electron clouds, also often referred to as a multipactor, is driven by the electric fields of successive bunches. This was first observed at the Intersecting Storage Rings (ISR, the world's first hadron collider built at CERN) in 1977 [38]. This multipactor arose from a resonance motion of a cloud of secondary electrons bouncing back and forth between opposite walls of the vacuum chamber. A consequence for the vacuum system was a strong electron stimulated gas desorption and an associated pressure increase, which affected the beam lifetime. This phenomenon is different from multipactor in RF structures but it shares many commonalities. Studies of electron clouds have developed quite extensive literature by themselves but this is outside the scope of our work.

Multipactor occurs in a localized part of an RF vacuum device, where the necessary conditions are satisfied, and may change its position when the electromagnetic field magnitude changes. Quite often, multipacting is encountered while raising the RF cavity field to the operating level, at much lower field levels than typically used for acceleration of particle beams.

Normal conducting cavities operating in a wide frequency range are often susceptible to multipacting. A couple of examples of such cavities are 52 MHz cavities for proton storage rings at DESY [39, 40] and 700 MHz cavity for the VEPP-5 booster ring at Budker INP [41]. Multipactor limits the cavity voltage to that of a particular barrier until it is processed away. The processing is performed in the pulsed mode first and then in the CW mode for many hours before cavity vacuum improves to an acceptable level and it becomes possible to increase the cavity voltage above the multipacting zone.

In the case of superconducting cavities, multipactor may prevent a cavity from reaching its operating field level and even lead to a local thermal breakdown (i.e., quench, or loss of superconductivity) due to excessive heat deposition onto the cavity surface. The SEY of cavity materials (niobium is the material predominantly used for SRF cavities nowadays) can be high enough to support multipacting even at low energies of primary electrons. This is due to an oblique angle of incidence caused by the magnetic field in some cases. Better understanding of conditions for multipactor existence can provide insights into how to modify structure geometries so that they are less susceptible to multipacting or even multipactor-free. Multipactor was a major factor limiting the performance of superconducting cavities for accelerators (Fig. 6.1) until a special elliptical shape of accelerating cavities was invented [42, 43] (see Fig. 6.2). Since then, multipactor rarely presents an insurmountable problem for structures accelerating near-speed-of-light particles.

Understanding the principle of minimal electric field at the multipactor location helps to avoid the appearance of this discharge in the transitions between beam pipes and cavities in accelerating structures, such as in the superconducting Ichiro cavity or in the Cornell ERL Injector cavity [44] (see Chap. 11). This can be done by a slight change of the geometry without compromising the main cavity parameters. One more geometrical solution is the introduction of rectangular and triangular grooves on surfaces prone to multipacting. This proved to be an efficient way to suppress multipactor in cathode inserts for photoemission SRF electron guns at

Brookhaven National Laboratory (BNL) [45] and at Helmholtz-Zentrum Dresden-Rossendorf (HZDR) [46].

Non-elliptically-shaped SRF cavities are used in accelerators as well, predominantly to accelerate particles with velocities much lower that the speed of light but sometimes even for ultra-relativistic particles if low-frequency structures are desired [47, 48]. Such structures are more susceptible to multipacting. Applying the shape modification method can help here as well. For example, initial consideration for a 56 MHz quarter-wave superconducting coaxial resonator [49] employed in the Relativistic Heavy Ion Collider (RHIC) at BNL indicated that it would suffer from severe multipacting. Introducing ripples on the outer conductor of the cavity allowed suppression of multipacting [50].

Multipactor can be overcome by a fast increase in voltage in the device so that there is not enough time for an electron avalanche to arise even though it has an exponential character. Mitigation is achieved through a strong coupling between the power source and the cavity, which makes the time for crossing the multipacting zone significantly shorter than the time constant of the avalanche. An example of utilizing such an approach for a 113 MHz SRF gun at BNL is given in [51]. If fast rise in voltage cannot be used, RF processing in CW or pulsed mode helps to clean the surface so that SEY decreases and multipacting disappears. Examples of RF processing setups can be found in the literature (see Section 8.9 in [52] and the references therein).

We will not discuss multipactor in the presence of permanent electric or magnetic fields because their usage in RF cavities is limited and presumably impossible in a superconducting RF cavity. For the same reason, we will not discuss multipactor in the presence of dielectrics.

There are works devoted to the characteristics of a strong multipactor when the multipactor current may reach a level comparable to the wall current of the cavity [53]. In this case the space charge contributes significantly to the dynamics of the electron flow and to the power losses. In a superconducting RF cavity, any additional losses will compromise its high quality factor and – if severe enough – cause quench, as we mentioned already. However, space charge dynamics does not affect the conditions under which multipactor emerges. The main goal of this book is to provide means of evaluating these conditions and hence predicting the emergence of multipactor and to give an insight into how to avoid multipacting by modifying the cavity geometry. Therefore, we shall deal only with the origin of the multipactor discharge when consideration of space charge or other saturation effects is not relevant.

Numerical simulations play an important role in multipactor studies. There are many 2D and 3D software packages developed within the scientific community as well as commercially. Some tools for multipacting simulations are reviewed in [54, 55] and also discussed in Chap. 5. 3D simulations are especially needed for input couplers and non-elliptical accelerating cavities. However, in some cases, an analytical solution could be a powerful means for understanding how to change the structure geometry to mitigate multipactor.

This book will provide more detailed overviews of different multipactor types with relevant references.

References

1. C. Gutton: Sur la décharge électrique à fréquence très élevée. *Comptes-Rendus Hebdomadaires des Séances de l'Académie des Sciences* **178**, 467 (1924)
2. C. et H. Gutton, Sur la décharge électrique en haute fréquence. Comptes-Rendus Hebdomadaires des Séances de l'Académie des Sciences **186**, 303 (1928)
3. H. Gutton, Recherches sur le propriétés diélectrique de gaz ionisés et la décharge en haute fréquence. Annales de physique **13**, 62 (1930)
4. P.T. Farnsworth, Television by electron image scanning. J. Franklin Inst. **2**, 411 (1934)
5. P.T. Farnsworth, U.S. Patent 2,071,517 (1937)
6. H. Alfvén, H.-J. Cohn-Peters, Eine neue Art von Hochfrequenz-Entladung im Vacuum und deren Verwendung als Ionenquelle. Arkiv för Matematik, Astronomi och Fysik **31A**, 18 (1944)
7. W.J. Gallagher, The multipactor electron gun. Proc. IEEE **57**(1), 94 (1969)
8. H. Bruining, *Physics and Applications of Secondary Electron Emission* (McGraw-Hill, New York, 1954)
9. R.A. Kishek et al., Multipactor discharge on metals and dielectrics: historical review and recent theories. Phys. Plasmas **5**, 5 (1998)
10. E.W.B. Gill, A. von Engel, Starting potentials of high-frequency gas discharge at low pressure. Proc. R. Soc. A **192**, 446 (1948)
11. A.J. Hatch, H.B. Williams, The secondary electron resonance mechanism of low-pressure high-frequency gas breakdown. J. Appl. Phys. **25**(4), 417 (1954)
12. A.J. Hatch, H.B. Williams, Multipacting modes of high frequency gaseous breakdown. Phys. Rev. **112**(4), 681 (1958)
13. J.R.M. Vaughan, Multipactor. IEEE Trans. Electron Devices **35**(7), 1172 (1988)
14. K. Krebs, H. Meerbach, Die Pendelvervielfachung von Sekundärelektronen. Ann. Phys. **15**, 189 (1955)
15. V.D. Shemelin, Existence zones for multipactor discharge. Sov. Phys. Tech. Phys. **31**, 9 (1986)
16. J.R.M. Vaughan, A new formula for secondary emission yield. IEEE Trans. Electron Devices **36**(9), 1963 (1989)
17. M.A. Furman, M.T.F. Pivi, Probabilistic model for the simulation of secondary electron emission. Phys. Rev. ST Accel. Beams **5**, 124404 (2002)
18. A. Kryazhev et al., Hybrid resonant modes of two-sided multipactor and transition to the polyphase regime. Phys. Plasmas **9**, 11 (2002)
19. V. Shemelin, Generalized phase stability in multipacting. Phys. Rev. ST Accel. Beams **14**, 092002 (2011)
20. V. Semenov et al., Multiphase regimes of single-surface multipactor. Phys. Plasmas **12**, 073508 (2005)
21. S. Riyopoulos, D. Chernin, D. Dialetis, Theory of electron multipactor in crossed fields. Phys. Plasmas **2**, 8 (1995)
22. V. Shemelin, Multipacting in crossed RF fields near cavity equator, in *Proceedings of EPAC 2004, European Particle Accelerator Conference*, Lucerne, 2004, p. 1075
23. V. Shemelin, Multipactor in crossed RF fields on the cavity equator. Phys. Rev. ST Accel. Beams **16**, 012002 (2013)
24. V. Semenov et al., Multipactor suppression in amplitude modulated radio frequency fields. Phys. Plasmas **8**(11), 5034 (2001)
25. S. Belomestnykh, Overview of input power coupler developments, pulsed and CW, in *Proceedings of 13th International Workshop on RF Superconductivity*, Beijing, 2007, p. 419

26. T. Abe et al., Multipactoring suppression by fine grooving of conductor surfaces of coaxial-line input couplers for high beam current storage rings. Phys. Rev. ST Accel. Beams **13**, 102001 (2010)
27. H. Padamsee, J. Knobloch, T. Hays, *RF Superconductivity for Accelerators* (Wiley, New York, 1998). ISBN 0-471-15432-6
28. E. Somersalo, P. Yla-Oijala, D. Proch, Analysis of multipacting in coaxial lines. In *Proceedings of PAC 1995, Particle Accelerator Conference*, Dallas, 1995, p. 1500
29. R. Udiljak et al., Multipactor in a coaxial transmission line. I. Analytical study. Phys. Plasmas **14**(3), 033508 (2007). V.E. Semenov et al., Multipactor in a coaxial transmission line. II. Particle-in-cell simulations. Phys. Plasmas **14**(3), 033509 (2007)
30. L.A. Kossyi et al., Experimental and numerical investigations of multipactor discharges in a coaxial waveguide. J. Phys. D: Appl. Phys. **43**, 345206 (2010)
31. R.L. Geng, H. Padamsee, Exploring multipacting characteristics of a rectangular waveguide, in *Proceedings of the 1999 Particle Accelerator Conference*, New York, 1999, p. 429
32. V.D. Shemelin, Multipactor discharge in a rectangular waveguide with regard to normal and tangential velocity components of secondary electrons. LNS Report SRF010322-03, Cornell University, Ithaca (2001). https://www.classe.cornell.edu/public/SRF/2001/SRF010322-03/SRF010322-03.pdf
33. R.L. Geng et al., Suppression of multipacting in rectangular coupler waveguides. Nucl. Instrum. Methods Phys. Res. A **508**, 227 (2003)
34. R.L. Geng et al., Dynamical aspects of multipacting induced discharge in a rectangular waveguide. Nucl. Instrum. Methods Phys. Res. A **538**, 189 (2005)
35. V.E. Semenov et al., Multipactor in rectangular waveguides. Phys. Plasmas **14**, 033501 (2007)
36. A.G. Sazontov, V.A. Sazontov, N.K. Vdovicheva, Multipactor breakdown prediction in a rectangular waveguide: statistical theory and simulation results. Contrib. Plasma Phys. **48**, 331 (2008)
37. C.J. Lingwood et al., Phase space analysis of multipactor saturation in rectangular waveguide. Phys. Plasmas **19**, 032106 (2012)
38. O. Gröbner, Beam induced multipacting, in *Proceedings of the 1997 Particle Accelerator Conference,* Vancouver, 1997, p. 3589
39. R.J. Burton, M.S. de Jong, L.W. Funk, Vacuum and multipactor performance of the HERA 52 MHz cavities, in *Proceedings of the 1990 European Particle Accelerator Conference,* Nice, 1990, p. 1017
40. I. Gonin et al., Multipactor phenomena in the 52 MHz PETRA II cavity at DESY, in *Proceedings of the 1994 European Particle Accelerator Conference,* London, 1994, p. 2197
41. K. Chernov et al., Startup of RF system for VEPP-5 booster ring, in *Proceedings of the XXth Russian Conference on Charged Particle Accelerators,* Novosiborsk, 2006, p. 188
42. U. Klein, D. Proch, Multipacting in superconducting RF structures, in Proceedings of Conference on Future Possibilities for Electron Accelerators, Charlottesville, 1979, p. N1
43. P. Kneisel, R. Vincon, J. Halbritter, First results on elliptically shaped cavities. Nucl. Intsrum. Methods Phys. Res. A **188**, 669 (1981)
44. S. Belomestnykh, V. Shemelin, Multipacting-free transitions between cavities and beam-pipes. Nucl. Instrum. Methods Phys. Res. A **595**, 293 (2008)
45. W. Xu et al., Multipacting-free quarter-wavelength choke joint design for BNL SRF, in *Proceedings of IPAC2015. International Particle Accelerator Conference*, Richmond, 2015, p. 1935
46. E.T. Tulu, U. van Rienen, A. Arnold, Systematic study of multipactor suppression techniques for a superconducting electron gun. Phys. Rev. Accel. Beams **21**, 113402 (2018)
47. M. Kelly, Superconducting radio-frequency cavities for low-beta particle accelerators. Rev. Accel. Sci. Technol. **5**, 185 (2012)
48. S. Belomestnykh, Superconducting radio-frequency systems for high-β particle accelerators. Rev. Accel. Sci. Technol. **5**, 147 (2012)
49. Q. Wu et al., Operation of the 56 MHz superconducting rf cavity in RHIC with higher order mode damper. Phys. Rev. Accel. Beams **22**, 102001 (2019)

50. D. Naik, I. Ben-Zvi, Suppressing multipacting in a 56 MHz quarter wave resonator. Phys. Rev. ST Accel. Beams **13**, 052001 (2010)
51. I. Petrushina et al., Mitigation of multipacting in 113 MHz superconducting rf photoinjector. Phys. Rev. Accel. Beams **21**, 082001 (2018)
52. H. Padamsee, *RF Superconductivity: Science, Technology and Applications* (Wiley-VCH, 2009). ISBN 978-3-527-40572-5
53. R. Kishek, Y. Y. Lau, Interaction of multipactor discharge and RF circuit. Phys. Rev. Lett. **75**(6), 1218 (1995)
54. F. Krawczyk, Status of multipacting simulation capabilities for SCRF applications, in *Proceedings of 10th Workshop on RF Superconductivity*, Tsukuba, 2001, p. 108
55. R.L. Geng, Multipacting simulations for superconducting cavities and RF coupler waveguides, in *Proceedings of PAC2003, Particle Accelerator Conference*, Portland, 2003, p. 264

Part I
Multipactor in a Planar Gap

Chapter 2
Existence Zones for Multipactor Discharge

2.1 Introduction

The first theories [2, 3] of multipactor discharge in a planar gap were based on an assumption that is not entirely consistent with available data on the energy distribution of secondary electrons. Specifically, to simplify the analysis, it was assumed that the velocity v_1 of secondary electrons is less than the velocity v_k of primary electrons by a constant factor $k = v_k/v_1$. The theory was later refined and a comparison was carried out with experimental results [4, 5]; however, this same assumption was used to calculate boundaries of zones in which discharge can exist.

In [6], the correctness of the simple model ($k = \text{const}$) was attributed to "natural selection", which ensures that only electrons with a fixed velocity ratio participate effectively in the growth of the discharge, and the distribution function for this ratio was found. However, the width of the distribution is quite large (k varies from 1.5 to 3), and the formulas in [3] and [4] indicate that the boundaries of the discharge zones are very sensitive to k in this range.

An alternative approach was considered in [7]. The initial velocity was assumed to be constant, and no allowance was made for a velocity distribution (according to the authors, this "would excessively complicate the problem"). A comparison with experimental results presented in [7] revealed a larger discrepancy than in [4]; however, the results were obtained with fewer assumptions.

The model constructed in [8] assumed that the secondary electrons have a small initial energy (2–3 eV) and allowed for their angular distribution. In addition, as in [7], the stability of electron motion in phase space and the possible return of electrons with a negative initial phase were analyzed in detail. The stability condition was analyzed without treating fluctuations in the initial velocity, and the

This chapter is based on material first published in [1].

V. D. Shemelin, S. A. Belomestnykh, *Multipactor in Accelerating Cavities*, Particle Acceleration and Detection, https://doi.org/10.1007/978-3-030-48198-8_2

normal components of the velocity were calculated assuming a Maxwell velocity distribution.

The experimental data [3–5] support the main assumptions of the theory, but they have a significant scatter, which is due primarily to errors in analyzing the data and the different surface materials that were studied. This may account for the tolerance of the experimental results to interpretation under assumptions that are physically not completely correct. These assumptions were not refined since they first appeared in papers published between 1954 and 1958 (e.g., see review in [9]) until [1], which is the subject of this chapter. The subsequent experimental and theoretical studies of secondary electron discharge have generally dealt with more complicated situations, such as discharges in superconducting cavities or on windows through which RF power is transmitted and the effects of magnetic field. In this chapter, we use experimental data on the spatial and energy distribution of electrons, including fluctuations in their initial velocity, and develop a systematic model for secondary electron RF discharge in a planar gap.

2.2 Distribution of Normal Velocity Components

As it crosses the gap, an electron acquires an energy much greater than its initial energy, and the tangential component of the final velocity can be neglected. For multipactor discharge, it is the component of the initial electron velocity normal to the electrode surface that is important.

The angular distribution of the secondary electrons is described by a polar diagram that is nearly circular, i.e., almost cosinusoidal, and the distribution is quite insensitive to the energy and incidence angle of primary electrons [10, 11] (see Fig. 2.1). We assume that the velocity distribution of secondary electrons has

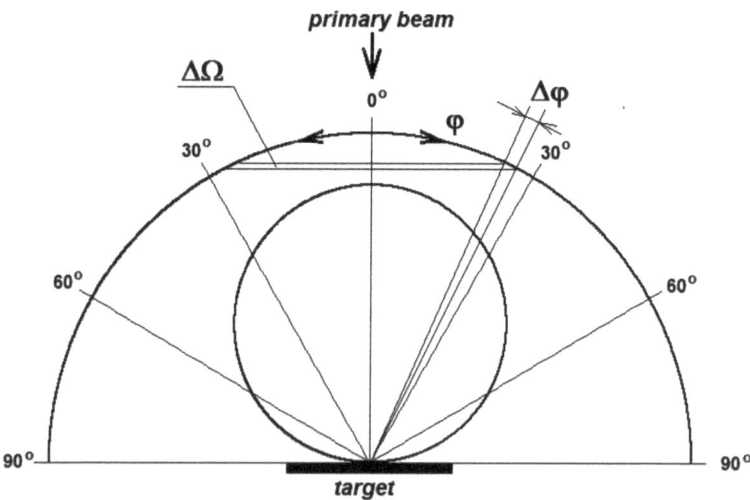

Fig. 2.1 Angular distribution of secondary electrons; somewhat simplified results from [11]

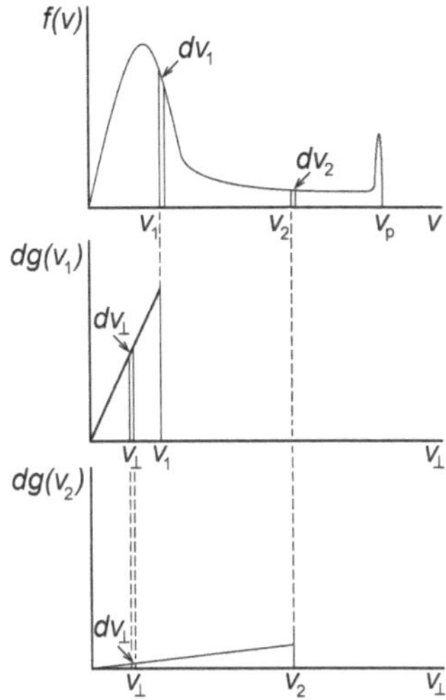

Fig. 2.2 Diagram pertaining to the calculation of the initial velocity distribution. The upper picture shows a typical distribution of secondary electrons with a broad peak of low-energy true secondary electrons and a narrow peak of back-scattered electrons. See also Fig. 2.3

been specified as $dN/N = f(v)dv$ (Fig. 2.2) and consider a group of n secondary electrons with velocities in a fixed interval $[v, v + dv]$. Their angular distribution relative to the normal to the surface is then given by $dn/n \propto \cos\varphi d\Omega$.[1] Here φ is the angle between the velocity vector of the secondary electrons and the normal to the surface, and $d\Omega$ is the element of solid angle. Because $d\Omega = 2\pi \sin\varphi\, d\varphi$, we have $dn/n \propto 2\pi \sin\varphi \cos\varphi\, d\varphi$. Applying the condition of normalization $\left(\frac{1}{n}\int dn = 1\right)$, we find that

$$dn/n = 2\sin\varphi\cos\varphi\, d\varphi.$$

At an angle φ, the transverse component of the velocity is $v_\perp = v\cos\varphi$ and its differential $dv_\perp = -v\sin\varphi d\varphi$. We can change variables as

$$\cos\varphi = v_\perp/v, \quad \sin\varphi d\varphi = -dv_\perp/v$$

to obtain

$$dn/n = 2v_\perp dv_\perp/v^2.$$

The minus sign has been omitted because it is determined by the direction in which φ is measured. Taking $dv_\perp > 0$ and recalling that $n = dN = Nf(v)dv$, we get

$$dn = \frac{2v_\perp dv_\perp}{v^2} Nf(v)dv. \tag{2.1}$$

[1]Following Bronshtein and Fraiman [12], we assume that the function $f(v)$ is independent of the angle φ, i.e., $F(v, \varphi) = f(v)g(\varphi)$.

We now fix $v_\perp$ and its differential $dv_\perp$ (Fig. 2.2) but let v vary. The numbers of electrons with different v (v_1 and v_2, say) but with $v_\perp$ lying in the same interval of width $dv_\perp$, are equal to

$$dn_1 = Nf(v_1)dv_1\frac{2v_\perp dv_\perp}{v_1^2}, \quad dn_2 = Nf(v_2)dv_2\frac{2v_\perp dv_\perp}{v_2^2}.$$

The areas of the triangles in Fig. 2.2 are equal to $f(v_1)dv_1$ and $f(v_2)dv_2$, respectively. Letting v vary from $v_\perp$ to v_p and integrating (2.1) over dv, we get the distribution

$$\frac{dN_1}{N} = 2v_\perp dv_\perp \int_{v_\perp}^{v_p} \frac{f(v)dv}{v^2} = g(v_\perp)dv_\perp$$

for the normal velocities. If we take $f(v)$ to be a δ–function, for example, we find that $g(v_\perp)$ is a linear function of $v_\perp$. For a narrow interval $dv_\perp$ contained in the interval $[v_\perp, v_p]$, we obtain

$$\left.\frac{dN_\perp}{N}\right|_{dv_1} = \begin{cases} 2v_\perp \dfrac{dN_1}{v_1^2} dv_\perp, & v_\perp \le v_1, \\ 0, & v_\perp > v_1. \end{cases}$$

The contribution from the interval dv_1 to the function $g(v_\perp)$ is thus described graphically by a triangle (Fig. 2.2), and the average value of the velocity for this contribution is $2v_1/3$. This result generalizes easily – the average value of the normal component is $2/3$ times the average of the absolute value of the velocity.

2.3 Analysis of Equation of Motion

The equation of motion for an electron in the gap is

$$\ddot{x} = \frac{e}{m}\frac{U}{d}\sin\omega t,$$

where the coordinate x is measured normal to the surface of one of the electrodes; e/m is the specific charge of the electron; U is the voltage across the gap, which is of width d; $\omega = 2\pi f$, where f is the oscillation frequency; and t is the time. The charge of the electron is assumed to be positive for simplicity of writing, so an electron starting at a negative phase is first decelerated. It is helpful to rewrite this in normalized form as follows [5, 7, 8]:

$$\lambda'' = \xi \sin\theta, \tag{2.2}$$

where $\lambda = x/d, \xi = U/U_0, U_0 = m\omega^2 d^2/e$, and $\theta = \omega t$; primes denote derivatives with respect to θ, while dots denote derivatives with respect to t.

Integrating Eq. (2.2), we obtain

$$\lambda' = \xi(\cos\theta_1 - \cos\theta) + \beta_1, \tag{2.3}$$

$$\lambda = \xi(\theta - \theta_1)\cos\theta_1 + \xi(\sin\theta_1 - \sin\theta) + \beta_1(\theta - \theta_1), \tag{2.4}$$

where θ_1 is the phase at which the electron enters the gap, and $\beta_1 = v_\perp/\omega d$ is the dimensionless normal component of the initial velocity of the secondary electron.

The condition for the electron to "resonantly" cross the gap is that the transit time be equal to an odd number of half-periods of the RF field. This ensures that newly generated secondary electrons see the same relative phase of the field as their predecessors. Equation (2.4) implies that

$$1 = \xi(\theta_2 - \theta_1)\cos\theta_1 + \xi(\sin\theta_1 - \sin\theta_2) + \beta_1(\theta_2 - \theta_1), \tag{2.5}$$

where θ_2 is the phase at which the electron reaches the second electrode at $\lambda = 1$. Because $\theta_2 - \theta_1 = (2n-1)\pi$, (here $n = 1,\ 2,\ \ldots$), equation (2.5) gives us

$$\xi = \frac{1 - (\theta_2 - \theta_1)\beta_1}{(\theta_2 - \theta_1)\cos\theta_1 + 2\sin\theta_1} = \frac{1 - (2n-1)\pi\beta_1}{(2n-1)\pi\cos\theta_1 + 2\sin\theta_1}. \tag{2.6}$$

The value of β_1 is given by

$$\beta_1 = \frac{\bar{v}_\perp}{\omega d} = \frac{2}{3}\frac{\bar{v}}{\omega d} = \frac{2}{3}\sqrt{\frac{2\overline{U_s}}{U_0}}, \tag{2.7}$$

where overlines denote average values, and $\overline{U_s}$ is the voltage corresponding to the mean initial energy $\overline{W_s} = e\overline{U_s}$ of the secondary electrons. It was found experimentally for copper that this mean energy is approximately 4 eV for a wide range of primary electron energies [12], as illustrated by Fig. 2.3.

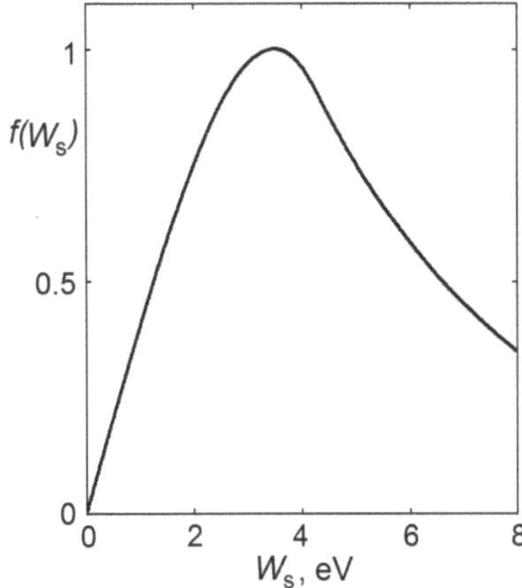

Fig. 2.3 The energy distribution function of secondary electrons for copper. The energy of primary electrons $W_p = 1.2$ keV. (Modified from [12])

2.4 Stability Condition

The condition for stable electron motion in multipactor discharge requires that the electron enters the gap at a definite phase. We can use formula (2.6) to calculate the normalized voltage ξ at which the discharge becomes unstable with respect to this phase.

The change in the phase at the exit from the gap is determined by fluctuations in the initial velocity and by the change in the phase at the entrance. This change is equal to the change in the phase for a secondary electron at the entrance to the gap:

$$d\theta_2 = a d\theta_1 + b d\beta_1.$$

Here the differentials represent physical deviations and are not, strictly speaking, infinitesimal. We take a and b to be derivatives:

$$a = \left.\frac{\partial\theta_2}{\partial\theta_1}\right|_{\theta_2-\theta_1=(2n-1)\pi} = \frac{(2n-1)\pi\xi\sin\theta_1 + \beta_1}{2\xi\cos\theta_1 + \beta_1},$$

$$b = \left.\frac{\partial\theta_2}{\partial\beta_1}\right|_{\theta_2-\theta_1=(2n-1)\pi} = -\frac{(2n-1)\pi}{2\xi\cos\theta_1 + \beta_1}.$$

which are calculated using (2.5).

After the gap has been crossed N times, the change in phase at the entrance is given by

$$d\theta_{N+1} = a^N d\theta_1 + b(a^{N-1} d\beta_1 + a^{N-2} d\beta_2 + \cdots + d\beta_N),$$

where the differentials $d\beta_1, d\beta_2, \ldots, d\beta_N$ are independent and correspond to fluctuations in the initial velocities. We write $d\beta_j = \alpha_j d\beta_0$ for $j = 1, 2, \ldots, N$, and for definiteness, we choose α_j and $d\beta_0$ such that $|\alpha_j| < 1$. Then,

$$d\theta_{N+1} = a^N d\theta_1 + b(a^{N-1}\alpha_1 + a^{N-2}\alpha_2 + \ldots \alpha_N) d\beta_0. \tag{2.8}$$

For $d\theta_N$ not to increase with N, the coefficients multiplying $d\theta_1$ and $d\beta_0$ must not increase with N. For the first coefficient, this requirement leads to

$$|a| = \left|\frac{(2n-1)\pi\sin\theta_1 + \beta_1/\xi}{2\cos\theta_1 + \beta_1/\xi}\right| < 1. \tag{2.9}$$

For the $d\beta_0$ coefficient to remain small, the variance of the sum in parentheses must not increase with N. Denoting the variance of α_j by σ_α^2, we find that the variance of the sum is

$$\sigma_s^2 = \sigma_\alpha^2 (a^{2N-2} + a^{2N-4} + \cdots + 1).$$

Because condition (2.9) bounds σ_s, the second term in (2.8) is also bounded when $|a| < 1$.

Because ξ in (2.6) increases as θ_1 decreases, for positive θ_1 in (2.9) we get the stability condition for the lower bound[2] of ξ:

$$\theta_1 < \theta_{1low} = \arctan \frac{2}{(2n-1)\pi}. \tag{2.10}$$

Substituting[3] (2.10) and (2.7) into (2.6) and expressing ξ in terms of U and U_0, we get the lower boundary of the multipactor discharge zone:

$$U_{low} = \frac{U_0 - (2n-1)\pi \cdot (2/3)\sqrt{2\overline{U}_s U_0}}{\sqrt{(2n-1)^2\pi^2 + 4}}.$$

On the upper boundary of the existence zone, $a = -1$. Solving this equation (with a taken from (2.9)) together with equation (2.6) yields the phase θ_1 bounding the zone from above:

$$\theta_1 > \theta_{1up} = -\arctan \frac{2}{(2n-1)\pi + \beta_1[4 - (2n-1)^2\pi^2]}. \tag{2.11}$$

Substituting (2.11) into (2.6) yields an upper bound for the voltage:

$$U_{up} = \frac{U_0 - (2n-1)\pi \cdot (2/3)\sqrt{2\overline{U}_s U_0}}{(2n-1)\pi \cos\theta_{1up} + 2\sin\theta_{1up}},$$

where θ_{1up} should be taken from (2.11).

We have thus found upper and lower bounds of the voltage for multipactor discharge to be stable against fluctuations in initial velocity. Figure 2.4[4] shows the corresponding boundaries, denoted by "1" and "2" for the fundamental multipactor zone $n = 1$. The upper part of the zone is shown on the right-hand side of the figure using the rightmost vertical scale.

[2] As it was noted by Vaughan [13], the expression (2.6) has its minimum at $\theta_1 = \theta_{1low}$ from (2.10). This is why he found the same lower limit for the multipactor zone as is found here. However, in his article it is not clear why θ_1 cannot be greater than θ_{1low} given that the "resonance" conditions are satisfied. Now the reason is clear – the condition of stability would be broken.

[3] Formula (2.10) is corrected from [1], where there is a typo.

[4] Lines for constant energy of primary electrons in Fig. 2.4 are corrected from [1].

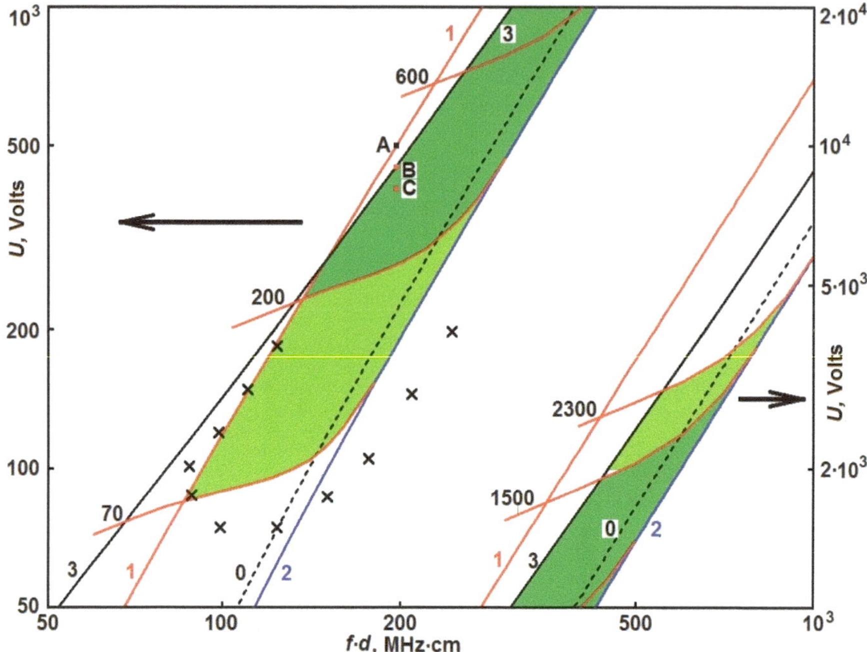

Fig. 2.4 Zone $n = 1$ for multipactor discharge. "1" and "2" indicate upper and lower stability bounds respectively; "3" indicates the bound determined by return of electrons to the electrode. The dotted line marked "0" corresponds to the starting phase $\theta_1 = 0$. The curves corresponding to constant primary electron energies are labeled with the corresponding values of energies: 70, 200, etc., in eV; $e\overline{U}_s = 4\,\text{eV}$. Points A, B, C, correspond to the curves in Fig. 2.5; the crosses are experimental values taken from [3]. The upper part of the zone is shown on the right-hand side of the figure with the corresponding vertical scale. An explanation of the discrepancy between the experimental and theoretical boundaries is given in Chap. 3

2.5 Returning Electrons

Particles emitted at negative RF phases experience decelerating force, turning them back toward the emitting plane. Most such particles are returned even though the "resonance" condition is satisfied. This gives ground to some authors [14] to examine only positive entrance phases. However, some particles, while having started at a negative phase, do not touch the surface of the emitting electrode and can still be accelerated to the opposite surface. It can be shown that multipactor zones become significantly wider if negative entrance phases are taken into account. The consideration of returning electrons imposes another restriction – beyond the stability conditions found in the previous section – in calculation of the multipactor zones' boundaries.

Figure 2.5 shows electron trajectories A and C for returning and non-returning electrons, respectively. The electron velocity vanishes at the outermost point of a

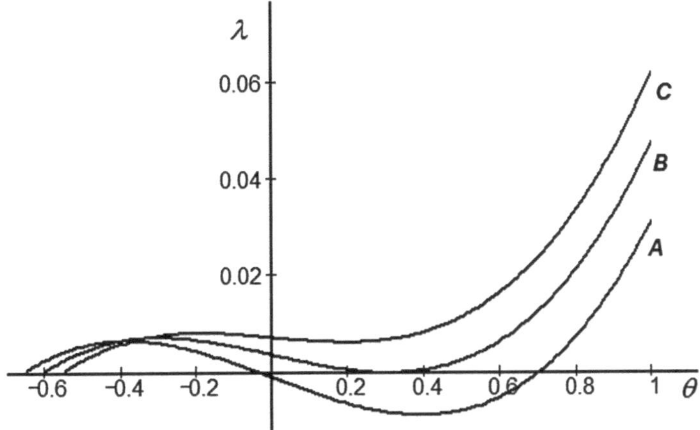

Fig. 2.5 Electron trajectories

returning trajectory. Equation (2.3) therefore gives us

$$0 = \xi(\cos\theta_1 - \cos\theta_{st}) + \beta_1, \tag{2.12}$$

where θ_{st} is the phase at which the electron stops. A trajectory represented by curve B in Fig. 2.5 touches the emitting surface and hence separates the returning electrons from the non-returning ones. In this case $\lambda = 0$ when $\theta = \theta_{st}$. Then equation (2.4) yields

$$0 = \xi(\theta_{st} - \theta_1)\cos\theta_1 + \xi(\sin\theta_1 - \sin\theta_{st}) + \beta_1(\theta_{st} - \theta_1). \tag{2.13}$$

Equations (2.12) and (2.13) form a system from which θ_{st} can be eliminated to find the critical value $(\beta_1/\xi)_{cr}$ such that for $\beta_1/\xi < (\beta_1/\xi)_{cr}$ the electrons return to the electrode from which they were emitted.

For a fixed value of fd, the bound imposed by the return of electrons with a negative initial phase can be found by solving (2.12), (2.13), and (2.6) simultaneously and setting $\beta_1/\xi = (\beta_1/\xi)_{cr}$. Figure 2.4 shows that on the one hand, for large fd, the returning electrons determine the boundary of the zone; in this case, the blocking voltage is large even for small negative phase at the entrance to the gap. On the other hand, for small fd, the blocking voltage is small and electrons can start out at large negative initial phases without returning; in this case, the stability condition determines the boundary of the zone.

A more precise analysis of the constraints due to returning electrons would require allowance for the fact that when the variance in β_1 is large, some of the electrons participating in the discharge return and thereby lower the effective SEY. When the SEY drops below unity, the discharge is extinguished. Here, we restrict ourselves to the case when the variance in β_1 is small.

2.6 Energy Constraints

For the discharge to grow, the energy of the incident electrons must be such that the SEY is greater than 1. A typical SEY curve, shown in Fig. 2.6, has three characteristic points. The first is a crossover point where SEY (δ) equals 1 at the corresponding energy E_1, then the point of maximum yield δ_{max}, which occurs at E_{max} energy, and finally the second crossover point where $\delta = 1$ again for E_2 energy. Therefore, no discharge will form in the gap if the gap voltage is too high ($>E_2$) or too low ($<E_1$), even if stable electron motion can exist between the electrodes.

The SEY depends strongly on the condition of the emitting surface. Let us consider copper, which is used in RF cavities and their auxiliary components. In [15], the authors reported that as-received copper can have a maximum SEY value as high as 2.5 at $E_p \approx 250\,\text{eV}$, while after sputter cleaning, the maximum value reduced to 1.3 and its energy shifted to $\approx$600 eV. Figure 2.7 illustrates the SEY of copper for two different surface conditions. Therefore, for a carefully cleaned copper surface, SEY > 1 in the energy interval 200–1500 eV; however, in practice, the copper surface is often oxidized, in which case this interval is enlarged [10, 12].

The energy bounds of a multipacting zone can be found from the equation

$$\beta_f = \sqrt{2U_f/U_0}, \tag{2.14}$$

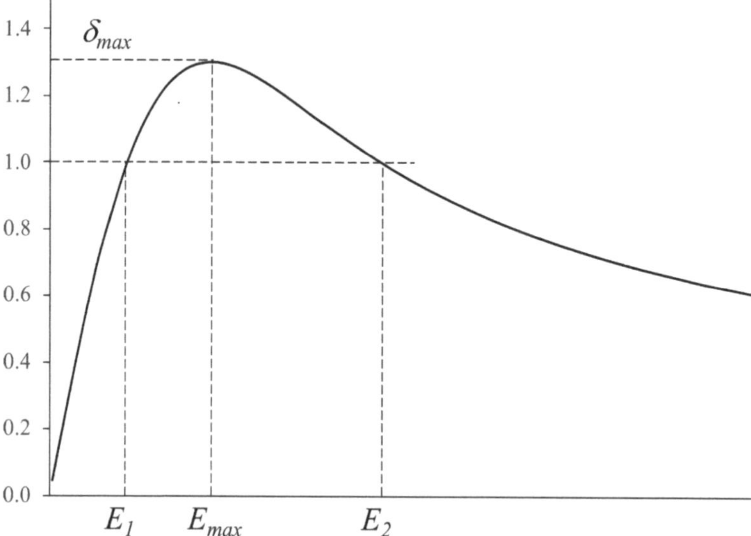

Fig. 2.6 Typical SEY curve as a function of primary electron energy

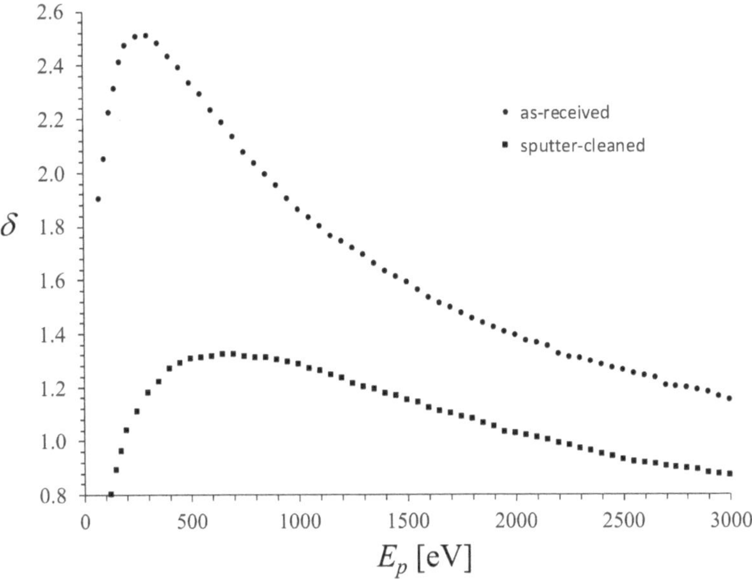

Fig. 2.7 SEY of copper; experimental data points taken from [15], Fig. 4

where β_f is the normalized electron velocity at the end of the trajectory ($\lambda = 1$), and $W_f = eU_f$ is the energy with which the electron reaches the electrode. Equation (2.3) gives

$$\beta_f = 2\xi \cos\theta_1 + \beta_1. \tag{2.15}$$

The bounds determined by the final energy of the electrons can be found by solving (2.6), (2.14), and (2.15) simultaneously with $U_f = 70$, 200, 1500, and 2300 V, as illustrated in Fig. 2.4. The curve $U_f = 600$ V corresponding to maximum SEY is also shown. This graph was constructed for the fundamental discharge zone $n = 1$. For comparison, the experimental data points [3] are shown in the figure as crosses.

2.7 Conclusion

In this chapter, we considered the zones of mutipactor discharge in a flat gap. Using experimental data on the angular and energy distributions of secondary electrons, we calculated bounds of existence for the discharge. There are several factors that determine the boundaries of multipactor zones. The basic condition for existence of multipactor is phase stability. Formulas for the phase stability bounds were derived. While most of the secondary electrons emitted at negative

RF phases are returning back to the emitting electrode, some do not touch the electrode and must be taken into account. The possible non-return of electrons with negative initial phases was considered, along with effects due to fluctuations in their initial velocity. These initial negative phases were shown to significantly expand the multipactor zone. Finally, this information in combination with SEY dependence on the primary electron energy was used to present the boundaries of the discharge zones analytically and graphically, and the results were compared with experimental data.

References

1. V.D. Shemelin, Existence zones for multipactor discharge. Sov. Phys. Tech. Phys. **31**, 9 (1986)
2. E.W.B. Gill, A. von Engel, Starting potentials of high-frequency gas discharge at low pressure. Proc. R. Soc. A **192**, 446 (1948)
3. A.J. Hatch, H.B. Williams, The secondary electron resonance mechanism of low-pressure high-frequency gas breakdown. J. Appl. Phys. **25**(4), 417 (1954)
4. A.J. Hatch, H.B. Williams, Multipacting modes of high frequency gaseous breakdown. Phys. Rev. **112**(4), 681 (1958)
5. B.A. Zager, V.G. Tishin, Resonant RF discharges and ways to suppress them. Zh. Tekh. Phys. **34**, 297 (1964) [Sov. Phys. Tech. Phys. **9**, 234 (1964)]
6. A. Miller, H.B. Williams, O. Theimer, Secondary-electron-emission phase-angle distribution in high-frequency multipacting discharge. J. Appl. Phys. **34**, 1673 (1963)
7. K. Krebs, H. Meerbach, Die Pendelvervielfachung von Sekundärelektronen. Ann. Phys. **15**, 189 (1955)
8. H. Tamagava, A theory of high frequency discharge in high vacuum. Electrotech. J. Japan. **3**, 93 (1957)
9. W.J. Gallagher, The multipactor effect. IEEE Trans. Nucl. Sci. **NS-26**, 4280 (1979)
10. H. Bruining, *Physics and Applications of Secondary Electron Emission* (McGraw-Hill, New York, 1954)
11. J.L.H. Jonker, The angular distribution of the secondary electrons of nickel. Philips Res. Rep. **6**, 372 (1951)
12. I.M. Bronshtein, V.S. Fraiman, *Secondary Electron Emission* [in Russian] (Nauka, Moscow, 1969)
13. J.R.M. Vaughan, Multipactor. IEEE Trans. Electron Dev. **35**(7), 1172 (1988)
14. P. Ylä-Oijala et al., Multipac 2.1 – Multipacting simulation toolbox with 2D FEM field solver and MATLAB graphical user interface. User's Manual, Rolf Nevanlinna Institute, Helsinki (2001)
15. I. Bojko, N. Hilleret, C. Scheuerlein, Influence of air exposures and thermal treatments on the secondary electron yield of copper. J. Vac. Sci. Technol. **A 18**, 972 (2000)

Chapter 3
Generalized Phase Stability in Multipacting

3.1 Introduction

The earlier theory of multipactor discharge in a flat gap [2, 3] explains the existence of boundaries of multipacting bands, or zones. Different zones correspond to different orders of discharge. The boundaries of the zones are related either to the condition of stability or to the minimal or maximal energy when the secondary emission yield becomes equal to 1. The literature on the subject of multipacting is quite extensive, but, as a rule, it generally deals with more complicated situations than a simple flat gap with homogeneous electric field. Nevertheless, even accurate analytical computations for this simple case of multipacting show some discrepancy with experimental results. Usually, one considers this discrepancy to be due to unknown details about the distribution of initial velocities of secondary electrons, some uncertainties in SEY, experimental errors, and so on. The influence of the normal and tangential components of the initial velocity on the position of multipacting zones was analyzed in Chap. 1 following [4] and [5]. However, the difference between calculated and measured boundaries has remained significant.

To highlight such a discrepancy, let us analyze the experimental results of a pioneer paper by A. Hatch and H. Williams [2] taking into account conditions of stability and limitations due to low impact energy. In the original work, the condition of stability was presented for a motion that looked like reflection from the walls with a certain elasticity coefficient. However, such a reflection is not supported by experimental data. The coefficient was just used to fit the experiment. Let us consider three points within the experimentally obtained multipacting zone: A, B, and C, as shown in Fig. 3.1. The theory presented in Chap. 2 explains the existence of the discharge only at point A. In this chapter, we present generalization of the

This chapter is based on material first published in [1].

V. D. Shemelin, S. A. Belomestnykh, *Multipactor in Accelerating Cavities*, Particle Acceleration and Detection, https://doi.org/10.1007/978-3-030-48198-8_3

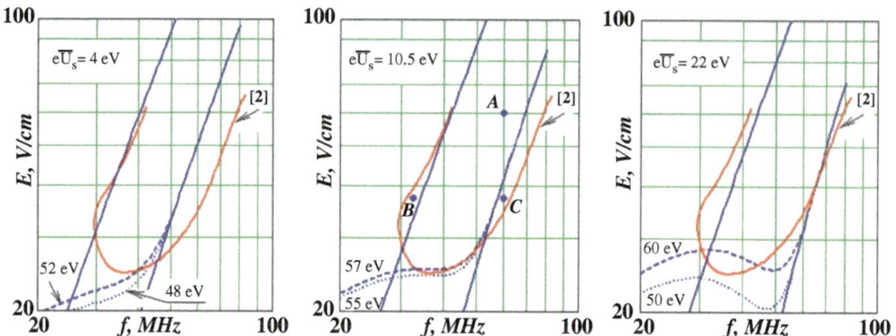

Fig. 3.1 Red: Experimental curve [2] and calculated boundaries of the multipactor zone for different values of initial energy of secondary electrons $e\overline{U}_s$. Blue: Straight solid lines are obtained from the condition of stability, and dashed and dotted lines correspond to a boundary with SEY = 1 calculated for different impact energies

phase stability condition (following [1]), which provides a better understanding of the experimental results.

Earlier, using (2.9), we obtained the condition of phase stability

$$|\partial\theta_2/\partial\theta_1| < 1. \tag{3.1}$$

We will call this the *simple stability condition*. Later we will introduce the concept of a *generalized stability condition*. Analysis of electron dynamics at points A, B, and C is straightforward, but we need to pay attention to the value β_1. The most probable initial velocity of a secondary electron is usually associated with the peak of the electron energy distribution. However, this assumption is wrong, and because the initial energy $e\overline{U}_s$ defines the value β_1, we will discuss this first.

3.2 Distribution of Initial Velocities and the SEY = 1 Boundary

The energy distribution of secondary electrons for silver electrodes used in the experiment under discussion can be taken from [6]. The expression for the energy distribution function $dn_e = f(W)dW$ can be recalculated for initial velocities or, as is more convenient, for the variable $\sqrt{W}$:

$$dn_e = f(mv^2/2)mvdv = 2\sqrt{W}f(W)d\sqrt{W}.$$

As can be seen from Fig. 3.2, the distribution with respect to $\sqrt{W}$ has a well-defined maximum corresponding to the most probable velocity of the **"true"** secondary electrons.

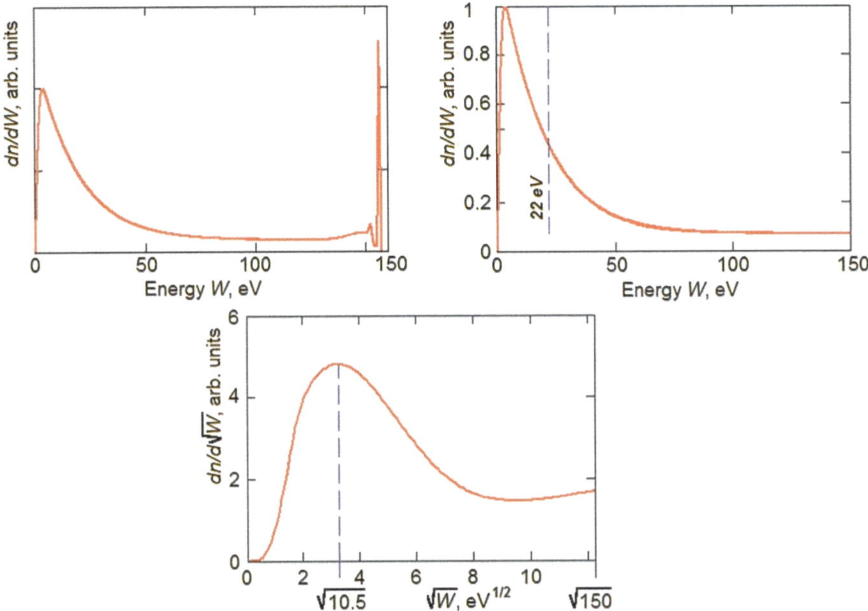

Fig. 3.2 Energy distribution of secondary electrons emitted by silver [6]; its analytical representation and distribution with respect to the square root of energy

For calculation of this distribution, the original experimental data (Fig. 3.2, upper left) were approximated analytically as

$$f(W) = 0.07 + p \cdot \exp\left(-\frac{W^l}{A}\right) - (p + 0.07) \cdot \exp\left(\frac{-W - 0.05W^2}{B}\right)$$

with $p = 1.15$, $l = 1.05$, $A = 22$, and $B = 3$. This gives an average value of 22 eV for the secondary electron energy, and the most probable velocity corresponds to an energy of 10.5 eV. Here, we neglect the elastically scattered electrons, responsible for the sharp maximum in the upper left of Fig. 3.2. In Fig. 3.1, the upper and lower boundaries of the first ($n = 1$) discharge zone are presented for different values of the initial energy. These boundaries are obtained in [4] from the condition (3.1). Also shown are the curves for the energy bounds corresponding to SEY = 1. One can see that the curvature of the line showing the energy bound fits experimental data best for an initial energy of 10.5 eV i.e. for the most probable velocity. The secondary electron initial energy of 4 eV used in earlier papers [4, 5] is not consistent with experimental results (see Fig. 3.1, left), though it does correspond to the maximum of the energy distribution. The mean value of secondary electron energy, 22 eV (Fig. 3.1, right), also does not fit the data well.

However, the initial energy value $e\overline{U_s} = 4\,\text{eV}$ appears to be correct if we analyze copper electrodes. Figure 3.3 (upper left) shows the experimental data [6]

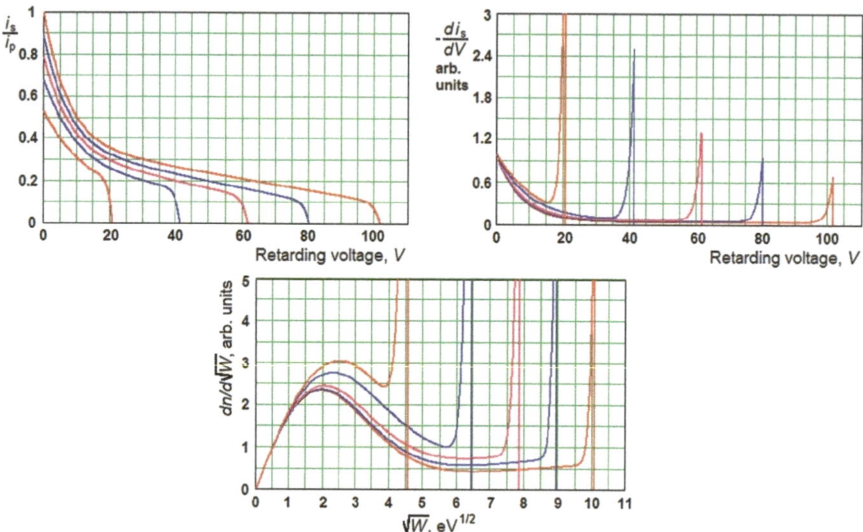

Fig. 3.3 Energy distribution of secondary electrons emitted by copper [6]; its analytically obtained derivative and distribution with respect to the square root of energy

for copper. The other two pictures in the figure present the mathematical handling of these data, similar to Fig. 3.2. It can be clearly seen that for the initial energy of primary electrons between 60 and 100 eV, the velocity peak position corresponds to secondary electron energies between 4 and 6 eV. These data for copper (Fig. 3.3, upper right) reveal a peak of the energy distribution at $V = 0$, not at 4 to 6 volts as it is for silver (Fig. 3.2, both upper pictures), and the peak of reflected electrons is too broad in comparison with the previous data for silver. Both these features are determined by derivatives at the ends of curves. However, even if there are some experimental errors at the ends, the peak of the velocity distribution at 4 to 6 eV ($\sqrt{W_s} = 2 \ldots 2.5$ eV$^{1/2}$, Fig. 3.3, lower) is determined by the middle part of these curves and can be treated as reliable.

Multipactor is a kinematic phenomenon, and now we can see that the defining value is the most probable velocity of the secondary electrons, whose position on the distribution curve does not necessarily coincide with the most probable energy.

3.3 Stability Condition for Different Points of the Multipacting Zone

The condition for stable electron motion in multipactor discharge requires that the electron enters the gap at a definite phase. We can use (2.6) to calculate the normalized voltage at which the discharge exists for a given start phase.

The change in the phase at the exit from the gap is determined by fluctuations in the initial velocity and by the initial change in the phase at the entrance. Here, we will neglect the initial velocity fluctuation because it is shown above (Sect. 2.4) that the same results are obtained if we take into account fluctuations in the initial phase only.

If the electron crosses the gap, the function $\theta_2 = f(\theta_1)$ can be obtained from (2.5). If the electron goes back to the same surface from which it was emitted, the value of 1 on the left-hand side of (2.5) should be replaced by 0.

Let us introduce a simple graphic interpretation of phase motion, focusing, and defocusing. With the help of a bisector of the right angle between coordinate axes, we can easily find the starting phase for the next gap-crossing if we know the previous starting phase and if the function $\theta_2 = f(\theta_1)$ is defined.

Examples of phase motion of an electron in the gap are shown in Fig. 3.4. Four different cases for the value used in (3.1) are shown in the picture. It is arbitrarily assumed for this figure that SEY = 1. Focusing (f) and defocusing (d) points are shown for positive (left picture) and negative (right picture) values of $\partial\theta_2/\partial\theta_1$.

Let us subtract an integer odd number of π radians from the phase θ_2 if the electron crosses the gap and subtract an even number of π radians from this phase if the electron falls on the same electrode it was emitted from. Let these integer numbers be such that the phases of arrival are in the interval $[-\pi, \pi]$. Then the phase of the particle "in resonance" will be the same after crossing the gap, i.e., $\theta_2 = \theta_1$. It is obvious that the condition of stability (3.1) will not change after such a transformation of θ_2. The advantage of this transformation lies in the fact that both phases, θ_1 and θ_2, are now placed on limited intervals, and the graphical interpretation of the phase motion becomes very illustrative.

Now, when we have defined the value of β_1 and redefined θ_2, we can return to construction of the function $\theta_2 = f(\theta_1)$ for points A, B, and C of Fig. 3.1. This function, the phase trajectories for an arbitrary initial starting phase, and the

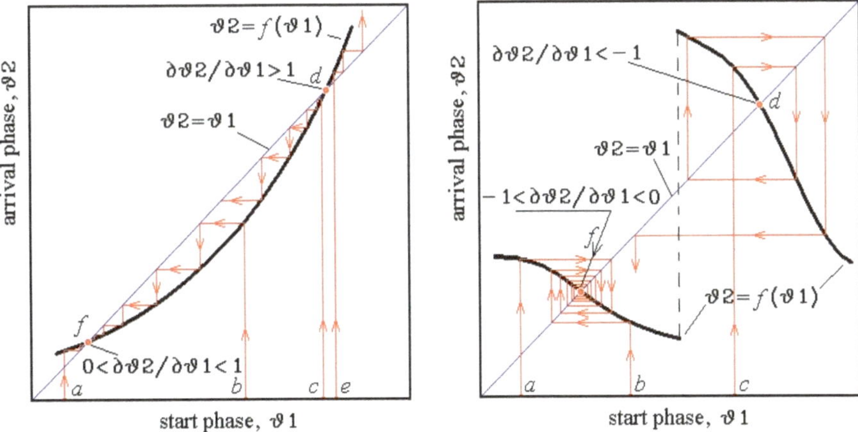

Fig. 3.4 Focusing to a stable phase and defocusing from an unstable phase for an increasing and decreasing function

corresponding impact energies are presented in Fig. 3.5. For each of the three points, 100 flights of the particle are calculated. It is assumed again that neither generation of new electrons nor loss of them occurs on the surface (SEY = 1).

The case of point A is very simple. The condition (3.1) is satisfied for $\theta_1 = -0.227$. After 5 crossings of the gap, the particle settles down to the focus point and all next phase positions coincide. Impact energy at this point is high enough, at approximately 130 eV (see the lower part of the picture), to produce secondary electrons with SEY > 1. Therefore, it is a point of multipacting.

The simple stability condition is not satisfied in point B because $\partial\theta_2/\partial\theta_1 = -1.09 < -1$. Let us distinguish two squares on the picture for this case. The inner one is tightly hatched because although a defocusing process develops, the value

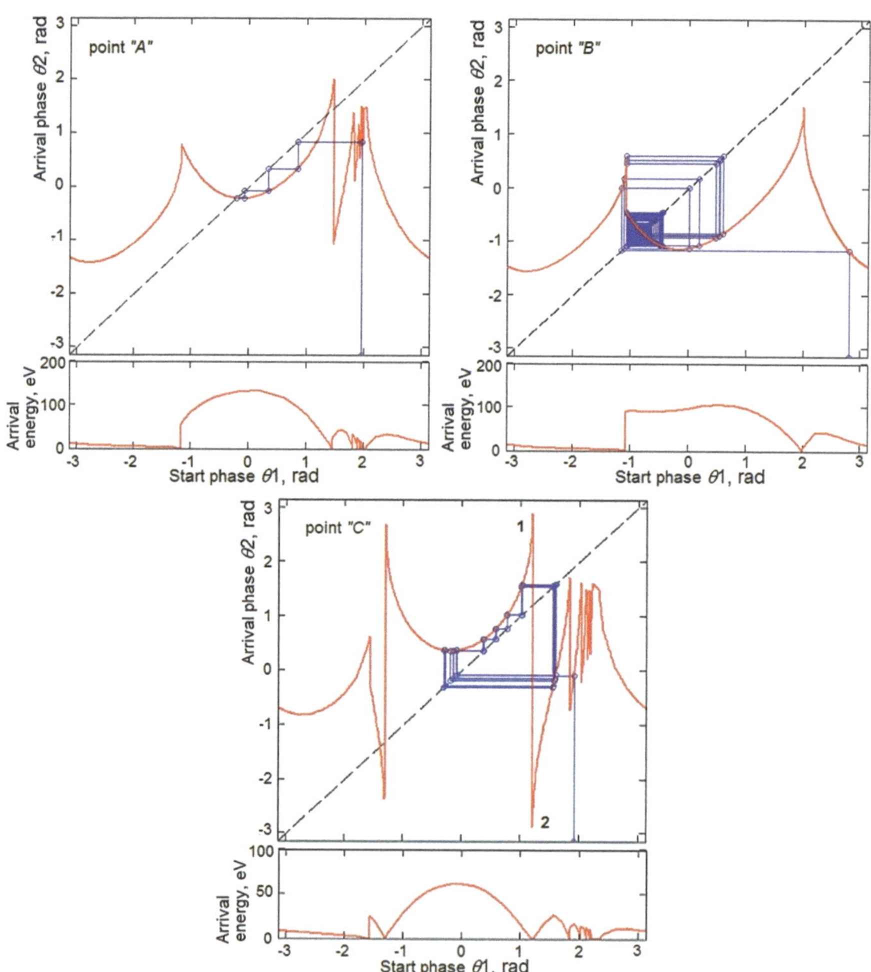

Fig. 3.5 Starting and arrival phases, and impact energy at arrival for the marked points in Fig. 3.1

of the derivative is not too high and lines go close one to another. The other one is the maximal square on the bisector the upper side of which is shown in Fig. 3.5 for point B ($MNPQ$ in Fig. 3.6a). The pattern of the phase trajectory practically does not depend on the initial start phase. When the phase trajectory reaches point M (see details in Fig. 3.6a), its continuation enters again into the smaller square. This trajectory cannot quit the square $MNPQ$, as can be seen from Fig. 3.6a, because after any small deviation of the trajectory outside the square, it enters again into this square. The necessary condition for this is that the entire curve is inside the square. In Fig. 3.6b, another possible behavior of the function $f(\theta_1)$ is shown: without a jump of the derivative at θ_1^l. We can formulate the first condition of the generalized stability as follows: if a part of the curve $\theta_2 = f(\theta_1)$ can be placed inside a square located on the bisector of coordinate axes as on the diagonal, with the values corresponding to the left and right sides of this square θ_1^l and θ_1^r, and if the related values θ_2^l and θ_2^r satisfy the ratio

$$\left|\frac{\theta_2^r - \theta_2^l}{\theta_1^r - \theta_1^l}\right| < 1, \tag{3.2}$$

the motion is finite and limited by phase angles θ_1^l and θ_1^r. The simple condition of stability (3.1), as can be easily seen, is a particular case of (3.2).

The whole curve defined on the interval $[-\pi, \pi]$ always satisfies condition (3.2), but does not make special physical sense and only presents another limiting case.

The case of point C is a case of repetitive motion when a part of the curve escapes from the square (peaks 1 and 2 in Fig. 3.5) but the whole trajectory stays within a limited interval. For description of this motion, let us introduce "higher order" functions:

$$\theta_3 = f_2(\theta_1) = f(f(\theta_1)), \theta_4 = f_3(\theta_1) = f(f_2(\theta 1)), \ldots$$
$$\theta_{m+1} = f_m(\theta_1) = f(f_{m-1}(\theta_1)). \tag{3.3}$$

It appears for the motion presented in Fig. 3.5, point C, that the equation

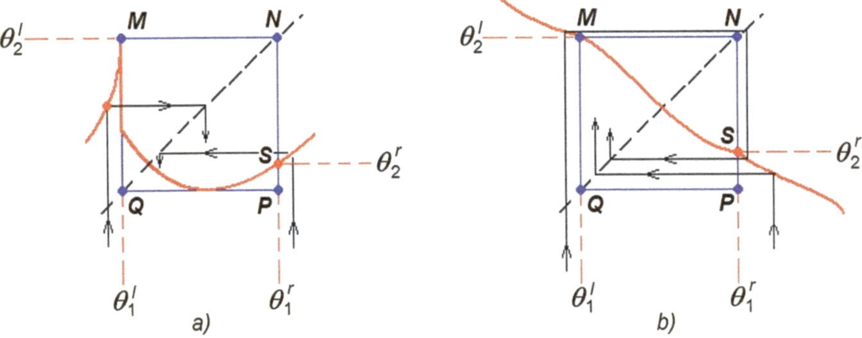

Fig. 3.6 Square of stability for point B

$$f_m(\theta_1) = \theta_1 \tag{3.4}$$

has a solution for $m = 6$. Therefore, the motion repeats after 6 different start phases. This motion could be stable if

$$|\partial f_m(\theta_1)/\partial\theta_1| < 1, \tag{3.5}$$

where $m = 6$ for point C. However, the inequality (3.5) is not satisfied in this case: $\partial\theta_7/\partial(\theta_1) = -3.86$. Nevertheless, because of the particular pattern of the phase trajectory, no particle falls into the region of peaks 1 and 2 (see Fig. 3.5, point C), and for the outlined square, the same condition (3.2) is valid as for point B.

The inequality (3.5) together with the condition (3.4) can be treated as the second generalized condition of stability. It reduces to the case of the simple stability condition in (3.1) when m is equal to 1. We should assume $f_1(\theta_1) \equiv f(\theta_1)$ in this case.

The lower parts of pictures in Fig. 3.5 show the energy of primary electrons versus the start phase. If this energy is too low, the discharge can survive only if the product of the SEY for all the cases of impact remains greater than unity. As can be seen from the data in [7] even at very low energies the value of SEY does not drop below approximately 0.7 due to elastically scattered electrons and the discharge can persist.

Another presentation of the phase motion is shown in Fig. 3.7, where the starting phase θ_m is plotted versus the flight number.

One more description of the grouping process is illustrated in Fig. 3.8, where for the initial phase θ_1 homogeneously distributed in the interval (1000 points), the next generations of particles are shown after different flight numbers. The two last pictures for points B and C show a change of phase for two successive flights.

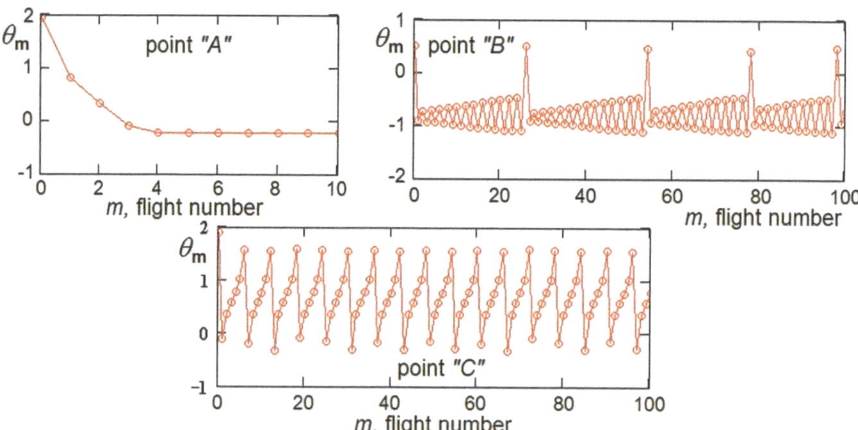

Fig. 3.7 Dependence of the starting phase on the flight number

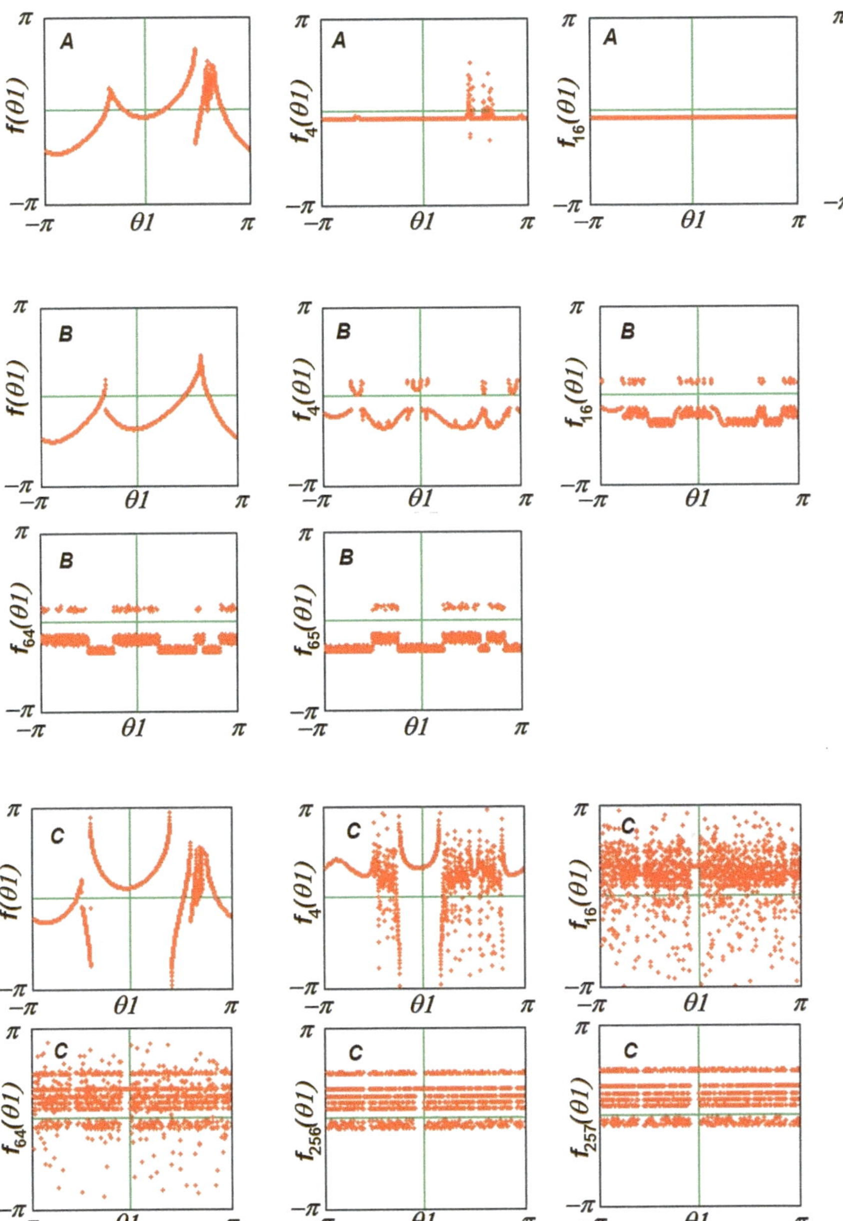

Fig. 3.8 Phase portrait of multipacting for points A, B, and C of Fig. 3.1 after different number of flights

3.4 Other Approaches to Phase Stability in a Flat Gap

In [8], for calculation of the phase stability, (3.5) was rewritten so that deviations from θ_1 and θ_2 are no longer infinitisemal. To obtain more exact equations, the cosine function in (3.5) was expanded to the second order and the sine function to the first order near the point of resonance. This procedure somehow increases the multipacting zone.

However, the procedure used in Sect. 3.3 includes any deviations from the resonance condition and covers all the results obtained in the cited paper.

3.5 Conclusion

Introduction of the generalized condition of stability in multipacting helped us to understand the expansion of multipacting zones beyond the limits predicted by the simple stability condition. The illustrative phase diagram for a flat gap showed the phase motion when the simple condition of stability is both applicable and not applicable for description of motion. Due to the assumption of SEY = 1, stable trajectories may not necessarily lead to growth, and the condition of cumulative SEY > 1 should be met in this model.

Other presentations of the grouping process showed that the electrons of multipactor discharge could group into layers even when the simple condition of stability did not work.

The results of this chapter are applicable to more complicated geometries when experimental zones appear broader than those obtained by simulations [9].

The results obtained for initial velocities of secondary electrons showed that the most probable initial velocity of secondary electrons does not correspond to their most probable energy, a fact that should be taken into account in simulations.

References

1. V. Shemelin, Generalized phase stability in multipacting. Phys. Rev. ST Accel. Beams **14**, 092002 (2011)
2. A.J. Hatch, H.B. Williams, The secondary electron resonance mechanism of low-pressure high-frequency gas breakdown. J. Appl. Phys. **25**(4), 417 (1954)
3. A.J. Hatch, H.B. Williams, Multipacting modes of high frequency gaseous breakdown. Phys. Rev. **112**(4), 681 (1958)
4. V.D. Shemelin, Existence zones for multipactor discharge. Sov. Phys. Tech. Phys. **31**, 9 (1986)
5. V.D. Shemelin, Multipactor discharge in a rectangular waveguide with regard to normal and tangential velocity components of secondary electrons. LNS Report SRF010322-03, Cornell University, Ithaca (2001). https://www.classe.cornell.edu/public/SRF/2001/SRF010322-03/SRF010322-03.pdf

6. H. Bruining, *Physics and Applications of Secondary Electron Emission* (McGraw-Hill, New York 1954)
7. J. de Lara et al., Multipactor prediction for on-board spacecraft RF equipment. IEEE Trans. Plasma Sci. **34**, 476 (2006)
8. M. Mostajeran, M.L. Rachti, On the phase stability in two-sided multipactor. Nucl. Instrum. Methods Phys. Res. A **615**, 1 (2010)
9. R.L. Geng, Multipacting simulations for superconducting cavities and RF coupler waveguides, in *Proceedings of PAC2003, Particle Accelerator Conference*, Portland (2003), p. 264

Chapter 4
Ping-Pong Modes

4.1 Introduction

Though more complicated modes of multipactor in a flat gap were discussed in the previous chapter, one of such modes has been studied in detail and named "ping-pong mode" in [2, 3]. In the ping-pong mode, the electron trajectory is divided into two parts during one half-period of the RF field. The first part of the trajectory occurs in the decelerating electric field, or at a negative starting phase, when the emitted electrons are returned to the emitting surface. Even though the impact energy is low, the returning electrons produce secondaries that follow the second part of the trajectory. The second part is mainly in the accelerating field, so that electrons propagate to the opposite electrode. An example of such a ping-pong trajectory is shown in Fig. 4.1.

According to the often used Vaughan's formula [4] there is a monotonous increase in SEY from zero to unity with increasing electron impact energy from zero to the so-called first crossover energy. However, it is known now that low-energy electrons can be completely reflected from the surfaces of different materials [5] (more references in [6]) with SEY only slightly less than 1. This is why we will consider the secondary electrons emerging after the first part of the trajectory as elastically reflected, i.e., having the same energy as the primary electrons.

The lowest mode of the ordinary two-surface multipactor has long been known and studied. Its upper voltage boundary also corresponds to negative starting phases and its lower boundary to positive ones. It has been argued that the area of existence of the ping-pong mode overlaps with the area of existence of ordinary two-sided multipactor bands.

Material of this chapter was first published in [1].

V. D. Shemelin, S. A. Belomestnykh, *Multipactor in Accelerating Cavities*, Particle Acceleration and Detection, https://doi.org/10.1007/978-3-030-48198-8_4

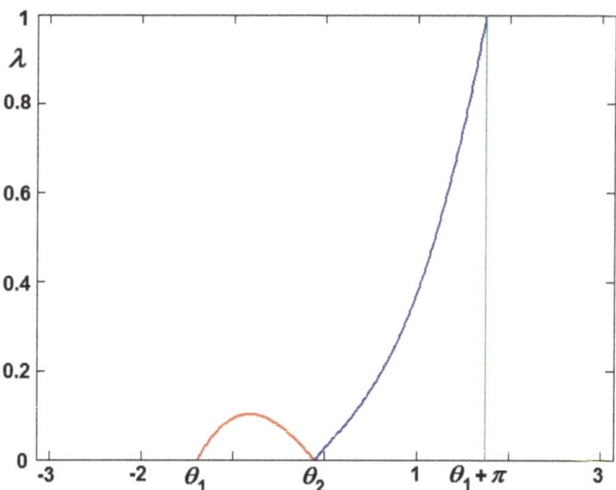

Fig. 4.1 An example of a particle trajectory in the ping-pong multipactor

In this chapter, we show that conditions of stability and cutoff limits for ping-pong modes are quite different from those derived in [2, 3] and that the overlapping area of the two multipactor bands is small.

4.2 Boundaries of Ping-Pong Modes

Equations of motion for an electron in a flat gap are given in Sect. 2.3, and they are the same for the ping-pong mode. Recall that the charge of an electron is taken as positive for simplisity of writing, so an electron starting at a negative phase is first decelerated.

4.2.1 *Stability Boundaries*

We will limit ourselves to the ping-pong multipactor of the first order. Equations of motion for the two parts of the electron trajectory and condition of stability $|d\theta_3/d\theta_1| < 1$ obtained in [2, 3] lead to the following equations for the stability boundaries:

$$\xi(\theta_2 - \theta_1)\cos\theta_1 + \xi(\sin\theta_1 - \sin\theta_2) + \beta_1(\theta_2 - \theta_1) = 0, \tag{4.1}$$

$$\xi(\pi + \theta_1 - \theta_2)\cos\theta_2 + \xi(\sin\theta_1 + \sin\theta_2) + \beta_1(\pi + \theta_1 - \theta_2) = 1, \tag{4.2}$$

$$\frac{\xi(\theta_2 - \theta_1)\sin\theta_1 + \beta_1}{\xi(\cos\theta_1 - \cos\theta_2) + \beta_1} \cdot \frac{\xi(\pi + \theta_1 - \theta_2)\sin\theta_2 + \beta_1}{\xi(\cos\theta_1 + \cos\theta_2) + \beta_1} = \pm 1, \tag{4.3}$$

where, θ_2 is the phase of the second start of an electron from the first surface and $\theta_3 = \pi + \theta_1$ is the phase of its arrival to the second surface (see Fig. 4.1).

In this analysis, it is assumed, as it is in [2] and [3], that secondary electrons start with the same velocity after each impact of the primary electrons on the surface. In the cited papers, this velocity corresponds to an energy of 3 eV. The stability boundaries of the normalized voltage according to equations (4.1)–(4.3) are shown in Fig. 4.2 as dashed lines. The upper limit M corresponds to the minus sign in (4.3), and the lower limit P corresponds to the plus sign. The line C represents the cutoff condition, which we will discuss in the next section.

From (2.3), we can find the velocity of the primary electron at the moment of impact θ_2:

$$\beta_2 = \xi(\cos\theta_1 - \cos\theta_2) + \beta_1. \tag{4.4}$$

In Fig. 4.3, the ratios β_2/β_1 are shown for both boundaries of stability, curves M and P.

We can see that the start velocity at the point θ_2 cannot be equal to β_1 because the energy of secondary electrons cannot be greater than the energy of primaries. As correctly stated in [3], "the distinction is poignant considering the higher proportion of back-scattered electrons for low primary impact energy". Therefore, let us consider that at the moment θ_2, the primary electrons reflect elastically and start with velocity $-\beta_2$.

Now, instead of equations (4.2) and (4.3) we should write

$$\begin{aligned}&\xi(\pi + \theta_1 - \theta_2)\cos\theta_2 + \xi(\sin\theta_2 + \sin\theta_1)\\&\quad - \xi(\cos\theta_1 - \cos\theta_2)(\pi + \theta_1 - \theta_2) - \beta_1(\pi + \theta_1 - \theta_2) = 1,\end{aligned} \tag{4.5}$$

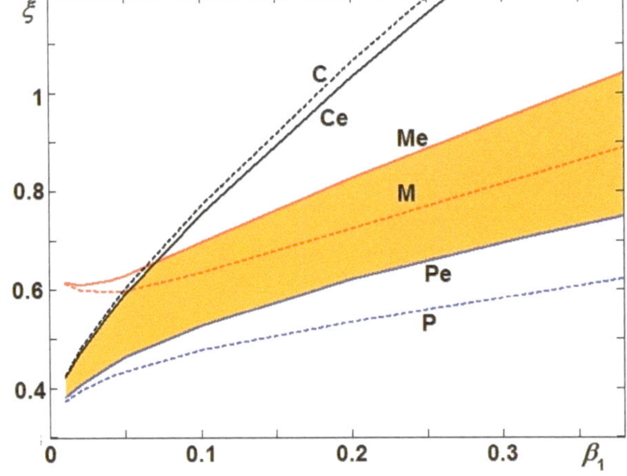

Fig. 4.2 Boundaries of the ping-pong multipactor defined by conditions of stability and cutoffs. See designations of the curves in text

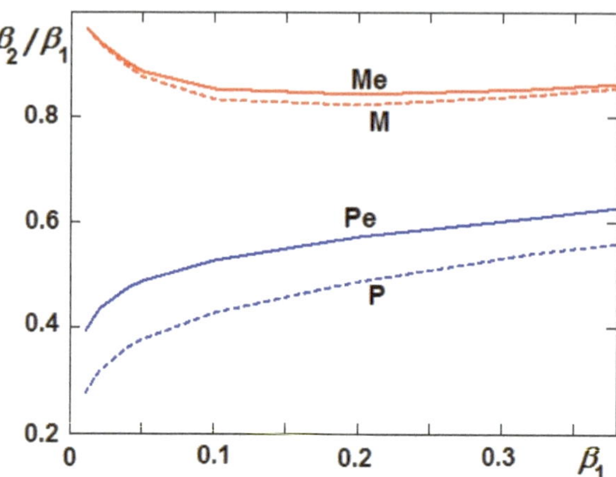

Fig. 4.3 Ratios β_2/β_1 for boundaries of stability

$$K - \frac{\xi(\pi + \theta_1 - \theta_2)\sin\theta_1}{2\xi\cos\theta_2 - \beta_1} = \pm 1, \tag{4.6}$$

where

$$K = [2\xi(\pi + \theta_1 - \theta_2)\sin\theta_2 - \xi(\cos\theta_1 - \cos\theta_2) - \beta_1] \cdot \frac{\xi(\theta_2 - \theta_1)\sin\theta_1 + \beta_1}{\xi(\cos\theta_1 - \cos\theta_2) + \beta_1}.$$

Solution of equations (4.1), (4.5), and (4.6) gives the stability boundaries presented in Fig. 4.2 as solid lines Me and Pe, where e denotes an elastic reflection at low energy. Ratios β_2/β_1 for the case of elastic reflection at the moment θ_2 are also shown in Fig. 4.3 as curves Pe and Me; they are still below unity as before.

4.2.2 Cutoff Boundaries

From solutions of the system (4.1), (4.5), (4.6), we should select only those that correspond to trajectories with $0 < \lambda < 1$. Therefore, we have cutoff conditions: $\lambda'(\theta_c) = 0$ and $\lambda(\theta_c) = 0$, where θ_c is the phase of the lowest point of the second part of the trajectory. These conditions correspond to the following equations:

$$\xi(2\cos\theta_2 - \cos\theta_c - \cos\theta_1) - \beta_1 = 0, \tag{4.7}$$

$$\xi(\theta_c - \theta_2)\cos\theta_2 + \xi(\sin\theta_2 - \sin\theta_c) - [\xi(\cos\theta_1 - \cos\theta_2) + \beta_1](\theta_c - \theta_2) = 0. \tag{4.8}$$

By solving these two equations together with (4.1) and (4.5) for different values of β_1, we obtain the cutoff boundary presented in Fig. 4.2 as Ce. The cutoff

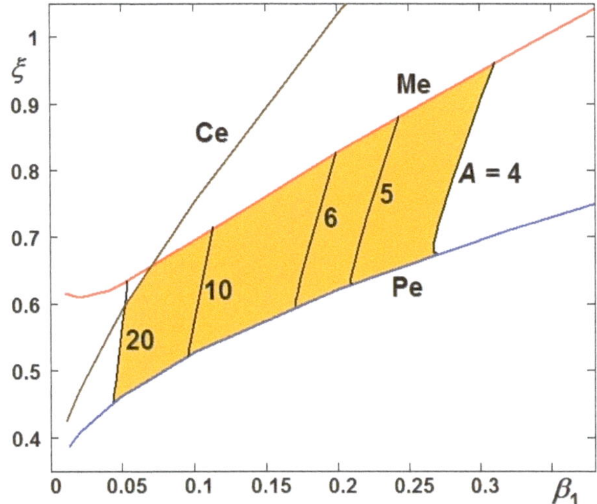

Fig. 4.4 Boundaries of stability (Me and Pe), cutoff (Ce), and lines of equal impact energy for the ping-pong multipactor

boundary for the case considered in [2, 3] (3 eV secondary electron energy for each impact) is shown as C.

4.2.3 Lines of Equal Impact Energy

After crossing the gap, electrons arrive at the other surface with velocity β_3. This velocity should be substantially higher than β_1 to compensate for a low SEY when they returned to the first surface with velocity β_2. Let us assume that $\beta_3 = A\beta_1$. Such an approach is convenient because it can be used for any initial energy of secondary electrons which in its turn depends on the material of the surface. In our case, we use β_1 corresponding to the starting energy of 3 eV as was done in [2, 3]. Then, the values $A = 1, 2, 3, \ldots, 20$ will correspond to 3, 12, 27, ..., 1200 eV.

In Fig. 4.4, we assume that SEY $\delta > 1$ when the energy of primary electrons $E_p > 48$ eV. The yellow-shaded area is the area of existence of the ping-pong multipactor. In fact, for the ping-pong multipactor, this energy should be much higher to compensate for the low energy at the point θ_2; we will use 48 eV as a reference energy.

4.3 Boundaries of the Two-Surface multipactor and Overlapping with the Ping-Pong Multipactor

Boundaries of stability and cutoff for the two-sided multipactor were found in earlier publications [7–9]. They are obtained here using the same approach as for the ping-pong multipactor. The lowest band of the two-sided multipactor according to this theory is presented in Fig. 4.5. Curves of equal energy of primary electrons are shown as those shown above for the ping-pong multipactor.

The united picture for both ping-pong and two-surface multipactor is presented in Fig. 4.6. The area where both kinds of multipactor can coexist is shaded in green. The lines of equal energy corresponding to $A = 20$ graphically coincide for both kinds of multipactor because they correspond to small β_1, and the trajectories in both cases are very close.

To transform to the conventional diagram of voltage versus fd, we use $U = \xi(m/e)\omega^2 d^2$ and $fd = v_1/(2\pi\beta)$, as illustrated in Fig. 4.7. Here, $v_1 = \sqrt{2(e/m)U_1}$, where eU_1 is the starting energy of secondary electrons at the point θ_1, taken as 3 eV. The left side of the ping-pong multipactor area is arbitrarily limited by arrival energy 48 eV. In reality, it should be much higher to make the SEY high enough to compensate for a low SEY at the point θ_2. This is

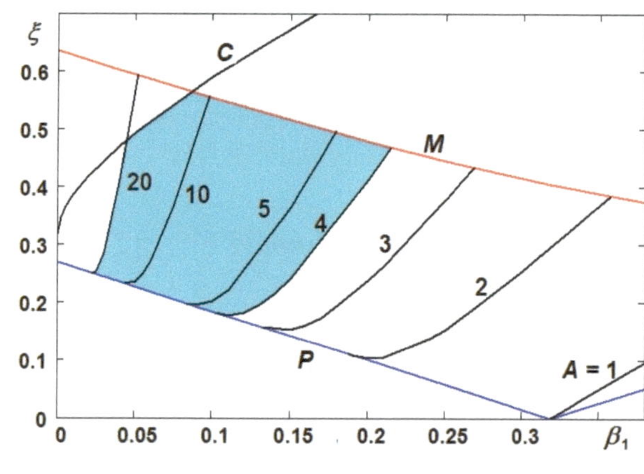

Fig. 4.5 Boundaries of stability, cutoff, and lines of equal impact energy for the two-surface multipactor

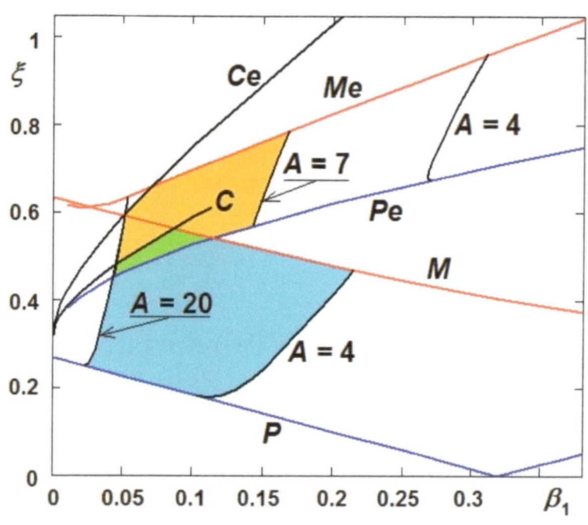

Fig. 4.6 United picture for the lowest bands of the two-surface and ping-pong multipactor

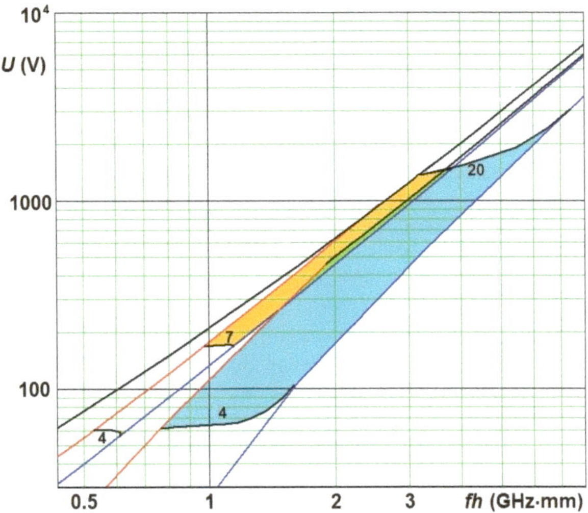

Fig. 4.7 Bands from Fig. 4.6 displayed as a conventional diagram of voltage vs fd

a possible reason why this area of multipactor is not observed experimentally. A continuous band of multipactor at low values of β_1, which is broader than given by this simplified analysis, is discussed in Chap. 3. It is explained by more than just multiple ping-pong single- and two-surface impacts per period.

The upper limit of voltage U is defined by the second crossover in SEY versus impact energy, when SEY becomes less than 1 for high impact energies; for clean copper, it is approximately 1500 eV or lower, down to 1000 eV [10]; in our graph $A = 20$ corresponds to 1200 eV.

4.4 Conclusion

In this chapter, we showed that the ping-pong mode of multipactor is located above the ordinary two-surface mode on the multipactor diagram (Fig. 4.7) because ping-pong mode electrons cannot start in an accelerating field. Therefore, the ping-pong mode cannot overlap both upper and lower boundaries of the two-surface multipactor, the opposite to what is stated in [2, 3]. In the cited papers, the author assumed that the secondary electrons are emitted with equal energies (3 eV) regardless of the starting phase. We have shown that the energy after the first part of the trajectory is less than the energy of the starting electrons, so only elastic reflection can lead to realization of the second part of the trajectory. The SEY for reflecting electrons can be close to 1 but still slightly lower. Thus, for the ping-pong multipactor to be sustainable, the SEY of the second electrode should be sufficiently high.

References

1. V. Shemelin, On boundaries of ping-pong modes in multipacting. Phys. Plasmas **25**, 053105 (2018)
2. R.A. Kishek, Ping-pong modes: a new form of multipactor. Phys. Rev. Lett. **108**, 035003 (2012)
3. R.A. Kishek, Ping-pong modes and higher periodicity multipactor. Phys. Plasmas **20**, 056702 (2013)
4. J.R.M. Vaughan, A new formula for secondary emission yield. IEEE Trans. Electron Devices **36**(9), 1963 (1989)
5. J. de Lara et al., Multipactor prediction for on-board spacecraft RF equipment. IEEE Trans. Plasma Sci. **34**, 476 (2006)
6. V.E. Semenov et al., Importance of reflection of low-energy electrons on multipactor susceptibility diagrams for narrow gaps. IEEE Trans. Plasma Sci. **37**(9), 1774 (2009)
7. K. Krebs, H. Meerbach, Die Pendelvervielfachung von Sekundärelektronen. Ann. Phys. **15**, 189 (1955)
8. A.J. Hatch, H.B. Williams, Multipacting modes of high frequency gaseous breakdown. Phys. Rev. **112**(4), 681 (1958)
9. V.D. Shemelin, Existence zones for multipactor discharge. Sov. Phys. Tech. Phys. **31**, 9 (1986)
10. V. Baglin et al., The secondary electron yield of technical materials and its variation with surface treatment, in *Proceedings of EPAC2000, European Particle Accelerator Conference*, Vienna (2000), p. 217

Chapter 5
Numerical Simulations of Multipactor

Numerical simulations play an important role in multipactor studies. However, in some cases, an analytical solution can be a powerful tool for understanding how to change the structure geometry to mitigate multipactor. We will demonstrate this in the following chapters.

The first application of computers to investigate multipacting can be traced back to the 1970s [1] when the phenomenon was a major SRF cavity performance limitation. Those computer simulations were instrumental in the discovery of a new type of mulitipactor, the one-point multipactor, by Lyneis et al. [2]. We will discuss one-point multipactor in detail in Chap. 9. Later, Klein and Proch developed the computer program MULTPAC and used it to study one- and two-point multipactor. Their studies resulted in proposing a spherical cavity shape as a way to avoid multipacting in SRF cavities [3]. We will consider two-point multipactor in the equatorial region of elliptical cavities in Chap. 8.

Nowadays, there are many 2D and 3D software packages developed within the scientific community as well as commercially. Some multipacting simulation codes are reviewed in [4, 5]. We can distinguish between general purpose codes that can deal with different geometries (2D or 3D) and ad hoc codes developed for specific cases.

Often these programs give results that are not in full agreement with each other and with experimental observations of multipactor. The results come close to experimental ones, as shown in [6], when a very dense mesh is used (approximately 2 millions cells for a 1300 MHz cavity) and the details of energy and angle distribution of all secondary electrons (true secondaries, rediffused and back-scattered)[1] are taken into account.

As the details of SEY become important for weak multipacting only, the analytical approach offers the ability to quickly sweep off unsatisfactory geometries at the initial stage of cavity design. Then, computer codes can be employed to

[1] The phase stability smooths the influence of velocity scatter of secondary electrons.

V. D. Shemelin, S. A. Belomestnykh, *Multipactor in Accelerating Cavities*, Particle Acceleration and Detection, https://doi.org/10.1007/978-3-030-48198-8_5

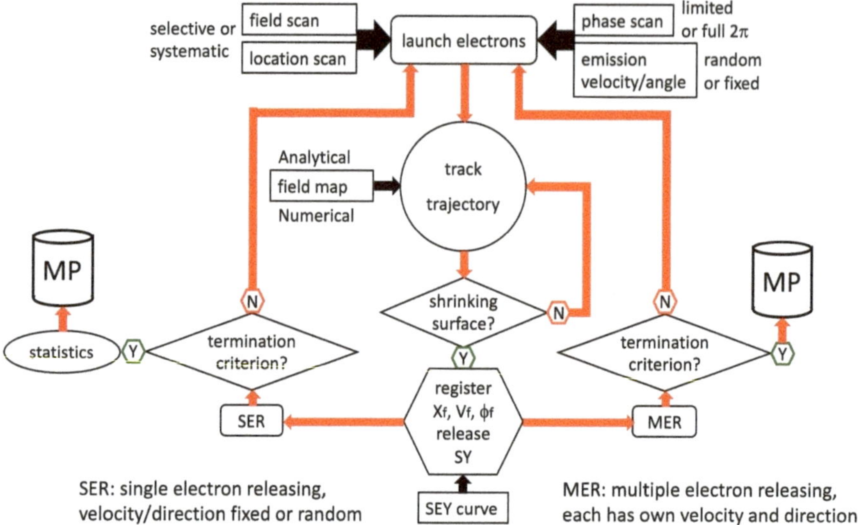

Fig. 5.1 Multipactor simulation algorithm [5]

finalize multipacting analysis. Any code for a complicated geometry can be and should be carefully benchmarked vs. a well-known geometry, such as a flat gap, and the results should be expected to be close to analytical solutions or experimental results [7].

The multipactor codes follow simulation algorithms similar to that shown in Fig. 5.1 [5]. For a given geometry, virtual electrons are launched at a selected field level from a selected surface location with respect to a chosen phase of RF field. The kinetic energy and direction of the electrons at emission are also specified. The trajectories of virtual electrons are tracked by solving equations of motion. After each integration step, it is checked whether the electron strikes a surface or not. If the answer is no, the integration goes on to the next step. If the answer is yes, the impact location, velocity of the electron, and phase of the RF field upon impact are registered. Then a new virtual electron is re-emitted from the impact site and tracking is continued until the next impact. After a certain number of impacts have been made, the tracking is stopped and an evaluation is performed to determine whether the multipacting conditions are satisfied or not. Different codes adopt either a single electron releasing scheme or multiple electron releasing scheme. A systematic scan of the field levels, electron emission locations, and RF phases is necessary to carefully identify boundaries of multipacting zones.

In the following, we consider examples of contemporary codes for multipactor simulations.

5.1 General Purpose Codes

In this section, we describe general purpose codes, one 2D and two 3D. General purpose codes are often "suites" of several codes designed to solve problems for a wide variety of RF structures. They may include several electromagnetic solvers in time and frequency domains, including interaction of particles with RF structures.

MultiPac is a 2D code developed at the University of Helsinki [8]. It is a simulation software toolbox for analyzing multipacting in axisymmetrical RF structures with the TM_{0nl}-mode, such as RF cavities, fundamental power couplers, and ceramic windows [9]. The software package includes a 2D finite element method (FEM) field solver. Examples of code outputs can be found in Chap. 8 (Fig. 8.1) and Chap. 11. MultiPac was very popular when it first appeared, but became less so with the development of 3D codes which allow simulations of more complicated geometries.

The code has one substantial drawback: particles cannot start at a negative phase, i.e., in a decelerating field. The authors write [8]: "The trajectory calculation is continued if the field phase is such that the possible secondary electrons are able to leave the wall." We believe that the secondary electrons can always leave the wall, but it is a different matter that they can return to the same surface. After returning, they can cross the gap, as in the case of ping-pong multipactor described in the previous chapter. The zone of existence of the simplest two-surface multipactor also becomes much broader if the electrons starting at a negative phase are taken into account. To summarize, we can say that if the multipactor is predicted by the MultiPac code, it will certainly exist, but it can also exist when the code doesn't predict it.

The suite of FEM electromagnetic codes ACE3P was developed at SLAC to be used on parallel computers to solve complex problems [10]. ACE3P consists of several modules, one of which is Track3P for multipacting in cavities and couplers and for studies of dark current in accelerating structures using particle tracking. Track3P was benchmarked with experiments. In particular, the code confirmed the existence of a hard barrier in the Ichiro cavity at approximately 30 MV/m with multipacting in the beam pipe tapered transition [11, 12] (see Chap. 11 for more details about this multipacting). Track3P finds wide use in the accelerator community to simulate multipacting not only in elliptical cavities but also in other cavity types [13] and in RF power couplers [14]. An example of particle trajectories obtained with Track3P for an SRF electron gun is shown in Fig. 5.2.

CST Studio Suite is another 3D package of electromagnetic simulation solvers often used for multipactor simulations. In particular, the Particle-in-Cell (PIC) solver is a versatile, self-consistent simulation method for particle tracking that calculates both particle trajectory and electromagnetic fields in the time-domain, taking into account space charge effects and mutual coupling between the two. This allows it to be used for simulations of multipacting [16]. The code uses an advanced probabilistic electron emission model by Furman and Pivi [17]. Figure 5.3 illustrates the output of CST Studio for multipacting in a reentrant cavity gap.

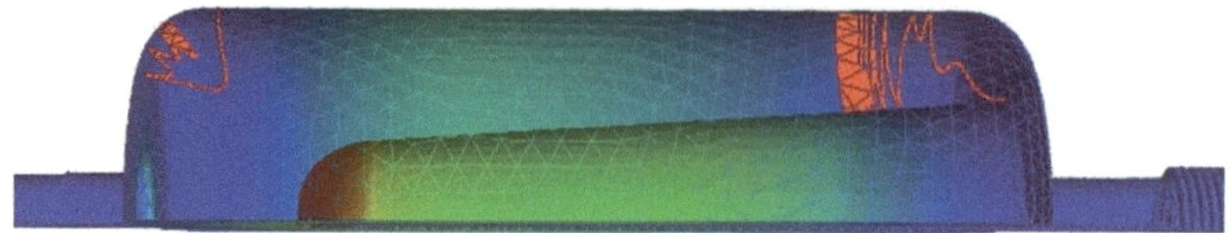

Fig. 5.2 Trajectories of electrons at two multipacting sites (red lines) in the 112 MHz SRF electron photoemission gun [15]. The left trajectory is for a gap voltage of 29 kV, while the right trajectory appears at a 40 kV gap voltage. Color indicates relative strength of the electric field

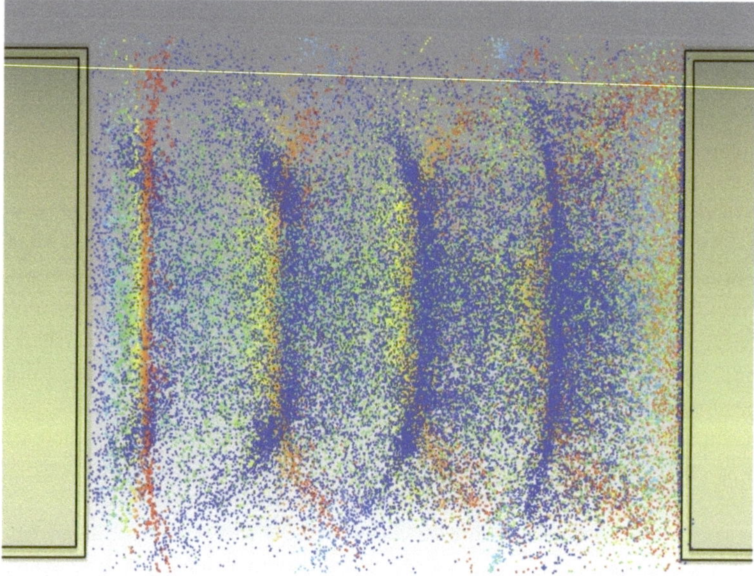

Fig. 5.3 6-order multipactor in an accelerating gap of a reentrant cavity (see Chap. 7) at 18.6 kV. This snapshot shows particles between two planes. Color indicates the particle energy from lowest (blue) to highest (red). Simulations were performed with CST Studio. (Courtesy of G. Romanov, Fermilab)

5.2 Ad Hoc Codes

Sometimes it is useful to write code to study a specific case. Such a code can provide more convenient outputs, simplify the problem to obtain crisper answers, and provide better insight into the specific problem being studied. Two examples of such ad hoc simulations are presented here.

In [18], Semenov et al. studied the influence of reflection of low-energy electrons on multipacting susceptibility diagrams. The authors conducted numerical simulations and found that the multiple reflections of low-energy electrons are responsible for significant broadening of the multipactor zone in narrow gaps. The starting point of this study is SEY $\delta \approx 1$ for very low energies of primary electrons and energy of secondary electrons close to the energy of primary ones. The same idea was used in the theory of the ping-pong mode (see Chap. 4), but here the number of reflections is not limited, and the particles are moving to the positive phase when they finally cross the gap and compensate low SEY gaining higher energy. The authors note that "... the absence of the left boundary of the first zone is not in agreement with the commonly accepted and experimentally observed picture of the multipactor susceptibility diagrams..." and suggested that this prediction should be studied experimentally.

A study of how the number of gap crossings can affect the assessment of the threshold value of the secondary yield was reported by Mostajeran and Rachti in [19]. The authors investigated a parallel plate model using the Monte Carlo method for numerical simulations and scanned a wide range of parameters. They showed that if the number of electron impacts is small (10–15), one can come to a wrong conclusion. Though no exact number of crossings was proposed, it was reported that 20–25 crossings lead to a better estimation of the multipactor threshold at the expense of a large simulation time.

References

1. I. Ben-Zvi, J.F. Crawford, J.P. Turneaure, Electron multiplication in cavities, in *Proceedings of PAC1973, Particle Accelerator Conference*, San Francisco (1973), p. 54
2. C.M. Lyneis, H.A. Schwettman, J.P. Turneaure, Elimination of electron multipacting in superconducting structures for electron accelerators. Appl. Phys. Lett. **1**(8), 541 (1977)
3. U. Klein, D. Proch, Multipacting in superconducting RF structures, in *Proceedings of Conference on Future Possibilities for Electron Accelerators*, Charlottesville (1979), p. N1
4. F. Krawczyk, Status of multipacting simulation capabilities for SCRF applications, in *Proceedings of 10th Workshop on RF Superconductivity*, Tsukuba (2001), p. 108
5. R.L. Geng, Multipacting simulations for superconducting cavities and RF coupler waveguides, in *Proceedings of PAC2003, Particle Accelerator Conference*, Portland (2003), p. 264
6. S. Kazakov, I. Gonin, V.P. Yakovlev, Multipactor simulation in SC elliptical shape cavities, in *Proceedings of IPAC2012, International Particle Accelerator Conference*, New Orleans (2012), p. 2327
7. G. Burt et al., Benchmarking simulations of multipactor in rectangular waveguides using CST–Particle Studio, in *Proceedings of SRF2009, International Conference on RF Superconductivity*, Berlin (2009), p. 321
8. P. Ylä-Oijala, D. Proch, MultiPac – Multipacting simulation package with 2D FEM field solver, in *Proceedings of the 10th Workshop on RF Superconductivity*, Tsukuba (2001), p. 105
9. P. Ylä-Oijala et al., *Multipac 2.1 – Multipacting simulation toolbox with 2D FEM field solver and MATLAB graphical user interface.* User's Manual, Rolf Nevanlinna Institute, Helsinki (2001)
10. K. Ko et al., Advances in parallel electromagnetic codes for accelerator science and development, in *Proceedings of LINAC2010, XXV Linear Accelerator Conference*, Tsukuba (2010), p. 1028
11. C. Ng et al., State of the art in EM field computation, in *Proceedings of EPAC2006, European Particle Accelerator Conference,* Edinburgh (2006), p. 2762

12. T. Saeki et al., Initial studies of 9-cell high-gradient superconducting cavities at KEK, in *Proceedings of LINAC2006, 2006 Linear Accelerator Conference*, Knoxville (2006), p. 794
13. I. Petrushina et al., Mitigation of multipacting in 113 MHz superconducting RF photoinjector. Phys. Rev. Accel. Beams **21**, 082001 (2018)
14. L. Ge et al., Multipacting simulations of TTF-III power coupler components, in *Proceedings of PAC2007, 2007 Particle Accelerator Conference*, Albuquerque (2007), p. 2436
15. T. Xin et al., Multipacting study of 112 MHz SRF electron gun, in *Proceedings of PAC2013, 2013 Particle Accelerator Conference*, Pasadena (2013), p. 1214
16. Electromagnetic Simulation Solvers CST Studio Suite. (Daussalt Systemes). https://www.3ds.com/products-services/simulia/products/cst-studio-suite/solvers/. Accessed 21 Nov 2019
17. M.A. Furman, M.T.F. Pivi, Probabilistic model for the simulation of secondary electron emission. Phys. Rev. ST Accel. Beams **5**, 124404 (2002)
18. V.E. Semenov et al., Importance of reflection of low-energy electrons on multipactor susceptibility diagrams for narrow gaps. IEEE Trans. Plasma Sci. **37**(9), 1774 (2009)
19. M. Mostajeran, M.L. Rachti, Importance of number of gap crossings on secondary emission in the simulation of two-sided multipactor. JINST **5**, P08003 (2010). http:/iopscience.iop.org/1748-0221/5/08/P08003

Part II
Multipactor in Crossed RF Fields

Chapter 6
Introduction of Accelerating RF Cavities

6.1 General Description

The most basic way to accelerate charged particles in particle accelerators is to use an electric field stored in radio frequency resonators, which are cavity resonators or simply RF cavities. In a particle accelerator, the cavity must have beam holes and beam pipes that allow a particle beam to pass through it. One of the simplest cavity shapes is cylindrical, or pillbox, as illustrated in Fig. 6.1a. Although the pillbox cavity geometry is never employed in accelerators (because of the absence of beam pipes), it is frequently used to illustrate figures of merit characterizing RF cavities (e.g., see [1]). Maxwell equations for the pillbox cavity can be solved analytically, resulting in an infinite number of solutions (eigenmodes). These eigenmodes belong to two classes with different field structures: transverse electric (TE) modes have a longitudinal component of the magnetic field while all components of the electric field are transverse with respect to the cavity axis z. In contrast, transverse magnetic (TM) modes have a longitudinal component of the electric field but only transverse components of the magnetic field. Only TM modes, which have the longitudinal component of the electric field, can accelerate particles. The lowest frequency mode, TM_{010}, of the pill-box cavity, or similar modes in other cavities, is used for acceleration of particles with velocities near the speed of light.

Contemporary RF cavities for particle acceleration are optimized for performance. Hence, their shapes deviate from the simple pillbox geometry and incorporate beam pipes. Normal conducting cavities often have a reentrant shape [2]. Figure 6.1b depicts a variant of the reentrant cavity called the nose-cone cavity. Superconducting RF cavities evolved into "elliptical cavities", which we will discuss in detail later in this chapter. A cavity can consist of one or more cells coupled together as shown in Fig. 6.2.

V. D. Shemelin, S. A. Belomestnykh, *Multipactor in Accelerating Cavities*, Particle Acceleration and Detection, https://doi.org/10.1007/978-3-030-48198-8_6

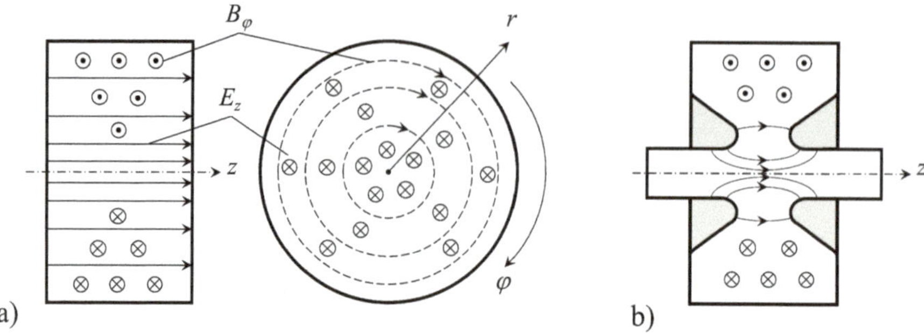

Fig. 6.1 (**a**) Pillbox cavity. (**b**) Nose-cone reentrant cavity with beam pipes

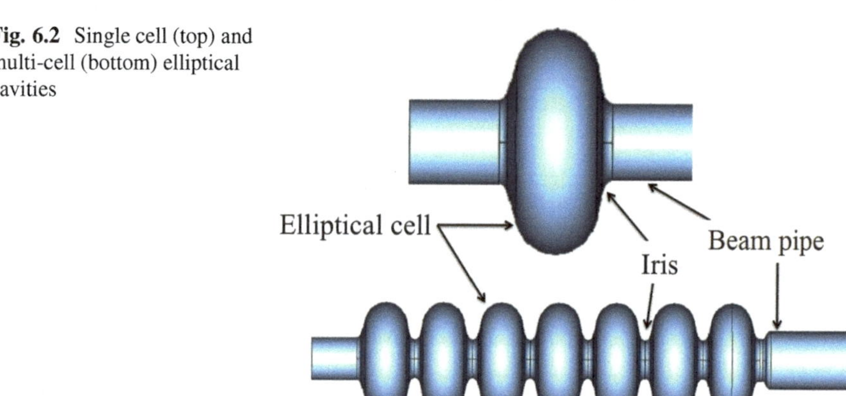

Fig. 6.2 Single cell (top) and multi-cell (bottom) elliptical cavities

6.2 Some Definitions

Consider a multi-cell cavity. The accelerating field E_{acc} in an inner cell is defined as the ratio of the energy gain ΔU, in volts, after the passage of a particle through the cavity to the length of the cell. For $\beta = 1$, this length ($2L$, Figs. 6.3 and 6.4) is equal to $\lambda/2$ or $L = \lambda/4$ (for π mode), where λ is the RF wave length. In the case of $\beta < 1$, we define the accelerating field as $E_{acc} = \Delta U/(\beta\lambda/2)$ because the optimal length of the half-cell is $L = \beta\lambda/4$ [1]. The particle velocity does not always correspond to the length of the cell. In this case, ΔU will differ from its optimal value calculated for the geometrical $\beta_{geo} = 4L/\lambda$. To avoid confusion, we will use E_{acc} corresponding to the geometrical beta.

The end cells do not have a physically defined exact length because the field propagates in the beam pipe exponentially decreasing (evanescent field). It appears logical for calculation of the accelerating field in the end cells to use the same definition: $E_{acc} = \Delta U/(\beta\lambda/2)$ so that there is no need to define the length of the end cells.

The peak surface electric and magnetic fields, E_{pk} and H_{pk}, play important roles in determining performance limits of superconducting cavities [1]. As the peak fields are proportional to E_{acc}, it is often convenient to use normalized peak fields E_{pk}/E_{acc} and H_{pk}/E_{acc}.

In Part III, we will consider optimization of the cavity shape to eliminate multipactor. A 1.3 GHz elliptical cavity will be used as a case study. It is convenient to compare different cavities obtained during optimization with a well known one. The TESLA cavity [3] is an obvious choice as it is widely used in various accelerators. For the peak surface electric field we will use a dimensionless parameter $e = E_{pk}/(2E_{acc})$ so that for the TESLA cavity $e = 1$. Analogously, the values of H_{pk}/E_{acc} are normalized to the corresponding value of TESLA, 42 Oe/(MV/m), so that $h = H_{pk}/E_{acc}/42$ is equal to 1 for TESLA cells.[1]

Other parameters used in the cavity optimization process are the geometric shunt impedance R_{sh}/Q and the geometry factor G. R_{sh} is the cavity shunt impedance and Q is the quality factor. To reduce the cavity wall losses, one must optimize the cavity shape for maximum $G \cdot R_{sh}/Q$. It is worth noting that the result of the cavity shape optimization does not depend on its frequency as all figures of merit – e, h, $G \cdot R_{sh}/Q$ – depend only on the geometry and not on the frequency.

6.3 Elliptical Cavity Geometry and Surface Fields

An elliptical cavity is a single- or multi-cell axially symmetric structure, similar to the cavities pictured in Fig. 6.2. The contour of a half-cell consists of two elliptical arcs and a straight segment tangential to both of them. The contour can be described by several geometrical parameters shown in Fig. 6.3 [6]. Three of these parameters, length of the half-cell L, aperture radius R_a, and equatorial radius R_{eq} are defined by physical requirements: L should be equal to a quarter of the wave length so that the particle can be effectively accelerated in each cell of a multicell cavity; the aperture is defined by requirements for coupling between cells and by the level of wake fields that can be allowed for a given accelerator; and the equatorial radius R_{eq} is used for tuning the cavity to a given frequency.

Of the remaining six parameters shown in Fig. 6.3 (A, B, a, b, d, γ), only four are needed to fully describe the geometry. Here A, B and a, b are the half-axes of the equatorial and iris constitutive ellipses, respectively. A combination of four parameters is selected for the cavity shape optimization.

Some authors follow [7] and use the following combination: $R = B/A$, the equator ellipse aspect ratio; $r = b/a$, the iris ellipse aspect ratio; the wall distance from the iris plane d; and the wall angle inclination γ. Thus, we have four dissimilar primary parameters: R, r, d, and γ. The reason for this choice is explained in [7] as

[1] According to our calculations, the normalized magnetic field is 41.4 Oe/(MV/m), approximately 1.5% less than this value. Publications cite this value as $H_{pk}/E_{acc} = 41.7$ Oe/(MV/m) [4] (Haebel, 1992), 42 Oe/(MV/m) [5] (Edwards, 1995), and 42.6 Oe/(MV/m) [3] (B. Aune et al., 2000). Normalization to 42 Oe/(MV/m) is chosen because (1) it is convenient to use a "round" number and (2) some publications refer to this value.

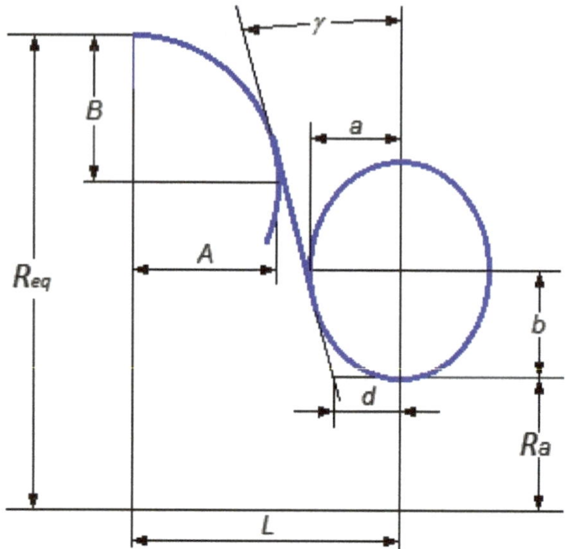

Fig. 6.3 Cell shape parameterization

follows: R defines a local minimum of the peak surface magnetic field, r defines a local minimum of the peak surface electric field, d allows a trade-off between the capacitive volume and the magnetic volume to balance the peak surface magnetic and electric fields on the cavity walls, and γ influences the mechanical stiffness of the cavity and controls its inductive volume.

This explanation looks convincing and physically feasible. However, a change of the angle γ or the permitted value of E_{pk}/E_{acc} leads to the necessity to optimize the values of R and r again. Besides, there is a dependence of the local minimum of the magnetic field on the iris aspect ratio though not so strong as on the equator aspect ratio, and conversely, the electric peak field minimum depends on the equator ellipse aspect ratio. On the other hand, the choice of four half-axes lengths (A, B, a, and b) as the primary parameters for optimization simplifies the description and does not complicate the optimization process. Recalculation from B/A, b/a, d, and γ to A, B, a and b can be easily done by solving a system of algebraic equations.

Equations defining half-axes of the smaller ellipse, a and b, and also coordinates of the point of conjugation (x_2, y_2) (point *x2* in Fig. 6.4), taking the values of γ, L, R_a, d, and r as given, are as follows:

$$\begin{aligned} \tan\gamma &= \frac{L-d-x_2}{y_2-R_a}, \\ \frac{(L-x_2)^2}{a^2} + \frac{(y_2-R_a-b)^2}{b^2} &= 1, \\ b(L-x_2)\tan\gamma &= a\sqrt{a^2-(L-x_2)^2}, \\ b/a &= r. \end{aligned} \tag{6.1}$$

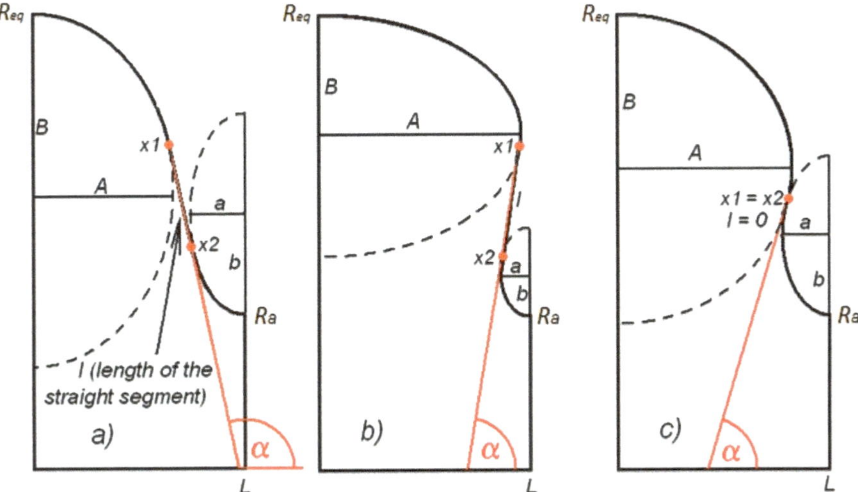

Fig. 6.4 Geometries of the elliptic half-cell: (**a**) non-reentrant and (**b**), (**c**) reentrant. Angle $\alpha = 90° + \gamma$, compare to Fig. 6.3

Half-axes of the bigger ellipse, A and B, and coordinates of the point of conjugation (x_1, y_1) (*x1* in Fig. 6.4), additionally taking the values of R_{eq} and R as given, are as follows:

$$\begin{aligned} \tan\gamma &= \frac{x_2 - x_1}{y_1 - y_2}, \\ \frac{x_1^2}{A^2} + \frac{(y_1 - R_{eq} + B)^2}{B^2} &= 1, \\ Bx_1 \tan\gamma &= A\sqrt{A^2 - x_1^2}, \\ B/A &= R. \end{aligned} \tag{6.2}$$

The half-axes A, B, a, and b are often used as a set of parameters for optimization, e.g., in [3, 8, 9]. We will use this set of primary geometric parameters further in this book.

Depending on the optimization constraints, one can obtain geometries from conventional non-reentrant to reentrant, as illustrated in Fig. 6.4. Here, $l = \sqrt{(x_2 - x_1)^2 + (y_2 - y_1)^2}$ is the length of the straight segment, and we measure the angle of the wall inclination between the axis of rotation and the straight segment of the wall as $\alpha = 90° + \gamma$, like it was done in our earlier publications. The cavity with $\alpha < 90°$ is known as the reentrant cavity.

Figure 6.5 compares three geometries: the TESLA cavity [3], the low-loss cavity [8] developed at JLab, and the Cornell reentrant cavity [9]. The surface electric and magnetic fields in these cavities are shown in Figs. 6.6 and 6.7. One can see that the

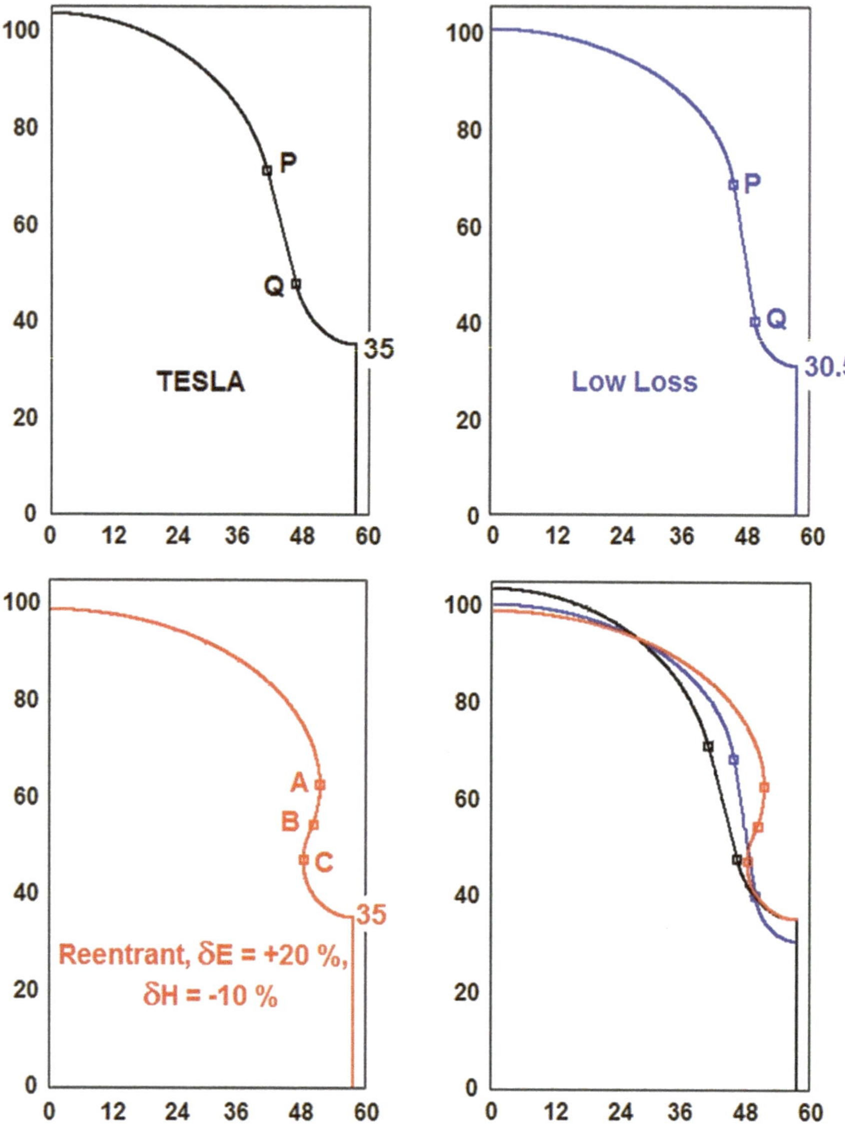

Fig. 6.5 Comparison of three known geometries of elliptic half-cells. All dimensions are in mm

electric field has irregularities at the points P and Q on the surface of non-reentrant cavities where the arc conjugates with the segment of the straight line. We believe that the best shape should have more regular dependencies for these curves, but for this purpose, we possibly have to abandon the paradigm of an elliptic shape and pass to a smoother contour, e.g., using some kind of splines [10].

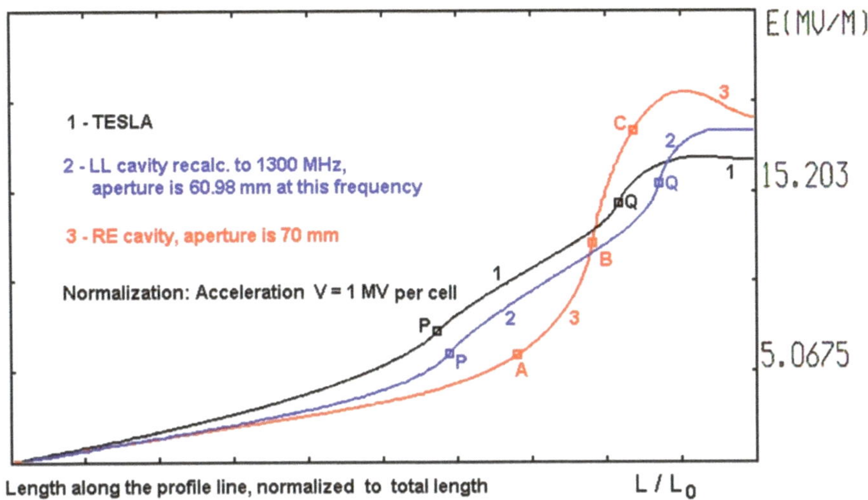

Fig. 6.6 Electric field on the surfaces of the cavities from Fig. 6.5

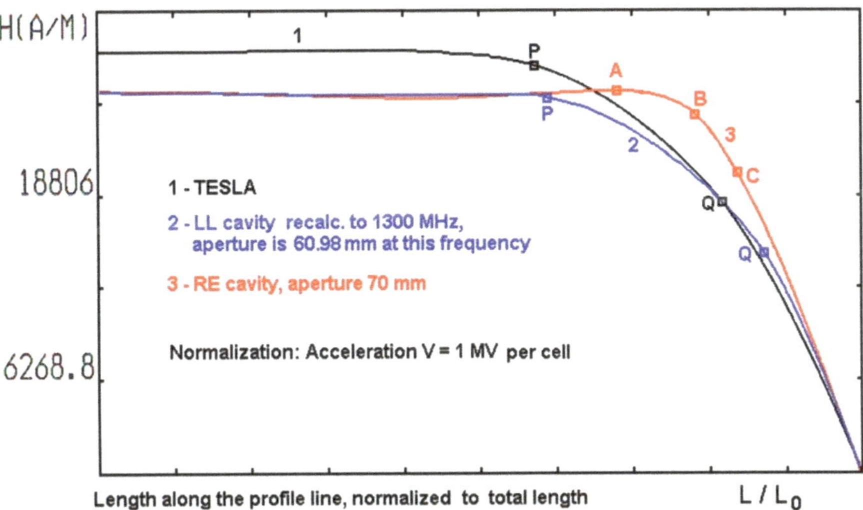

Fig. 6.7 Magnetic field on the surfaces of the cavities from Fig. 6.5

In the case of the reentrant cavity, the maximum of E is not flat as it is in non-reentrant geometry. One can suppose that flattening of this maximum will decrease the peak electric field but E_{acc} should not change significantly.

Comparing the magnetic fields of the TESLA cell and the reentrant cell having the same aperture, we can see that a decrease in the maximal field is obtained through the lengthening of the maximal field region. Again, better flattening of the maximal values of this field could decrease the peak field but only slightly. The area under these curves is approximately the same in both cases. In the case of the Low Loss cavity, the decrease in the field is obtained due to a lesser aperture. As shown in [11], H_{pk}/E_{acc} is a monotone function of the aperture radius, and it is smaller here due to a smaller R_a.

References

1. H. Padamsee, J. Knobloch, T. Hays, *RF Superconductivity for Accelerators* (Wiley, New York, 1998). ISBN 0-471-15432-6
2. M. Puglisi, Conventional RF cavity design, in *Proceedings of CERN Accelerator School on RF Engineering for Particle Accelerators*, Oxford, 1991, CERN publication 92-03 (1992), p. 156
3. B. Aune et al., Superconducting TESLA cavities. Phys. Rev. ST Accel. Beams **3**, 092001 (2000)
4. E. Haebel, A. Mosnier, J. Sekutowicz, Cavity shape optimization for a superconducting linear collider, in *Proceedings of the XV International Conference on HEACC*, vol. 2, Hamburg, 1992, p. 957
5. TESLA Test Facility Linac – Design Report. Editor D. A. Edwards. DESY Print, TESLA 95-01 (Mar 1995)
6. S. Belomestnykh, V. Shemelin, High-beta cavity design – a tutorial, in *Proceedings of the 12th International Workshop on RF Superconductivity* (Cornell University, Ithaca/New York, 2005) p. 2
7. P. Pierini et al., Cavity design tools and applications to the TRASCO project, in *Proceedings of 9th Workshop on RF Superconductivity*, Santa Fe, 1999, p. 380
8. J. Sekutowicz et al., Low loss cavity for the 12 GeV CEBAF Upgrade. JLab–TN–02–023 (June 2002)
9. V. Shemelin, H. Padamsee, R.L. Geng, Optimal cells for TESLA accelerating structure. Nucl. Inst. Methods Phys. Res. A **496**, 1 (2003)
10. B. Riemann, T. Weis, A. Neumann, Design of SRF cavities with cell profiles based on Bezier splines, in *Proceedings of ICAP2012, International Computational Accelerator Physics Conference*, Rostock-Warnemünde, 2012, p. 167
11. V. Shemelin et al., Systematical study on superconducting radio frequency elliptic cavity shapes applicable to future high energy accelerators and energy recovery linacs. Phys. Rev. Accel. Beams **19**, 102002 (2016)

Chapter 7
Effect of RF Cavity Magnetic Field on Multipactor in a Gap

7.1 Experimental Studies with 430 MHz Test Cavity[1]

During operation of the storage ring VEPP-3 [3], the voltage on the 72.5 MHz accelerating cavity is slowly varied through a wide range of amplitudes. At some voltage levels, this cavity can be prone to multipacting, which prevents an increase in the voltage.

It was necessary to study multipactor and find a means of suppressing it. An experimental setup was developed for this purpose. Keeping in mind that multipacting zones depend on fd and to ease the handling, a smaller test cavity with the geometry close to the VEPP-3 cavity but scaled to a higher frequency (430 MHz) was built. A simplified view of the experimental setup is shown in Fig. 7.1. The main dimensions of the cavity are illustrated in Fig. 7.2: the accelerating gap of the cavity $H - 2h = 125$ mm, the diameter $2a$ is 465 mm, the height of the reentrant stubs h is 48 mm, and the diameter of reentrant stubs $2b = 107$ mm. The quality factor of the cavity is $Q_0 = 20 \cdot 10^3$, and shunt impedance $R_{sh} = 3.4 \cdot 10^6$ Ohm. The copper cavity was enclosed in a stainless steel vacuum volume and pumped out to the vacuum of 10^{-6} Torr. Pulsed input power up to 10 kW was supplied from an RF generator with the rectangular pulse duration of 150 mcsec and repetition rate of 2.5 Hz. The voltage on the accelerating gap was calculated from the input and reflected power as $U_{res} = \sqrt{2(P_{in} - P_{refl})R_{sh}}$ and was additionally checked by an output loop with a calibrated detector.

The cavity also had an antimultipactor ring concentric with the axis and laying in the symmetry plane (not shown in Figs. 7.1 and 7.2). This ring was utilized to provide a high voltage bias used for suppression of multipactor and also for conditioning the inner surface of the cavity with glow discharge. We will not discuss

[1]Material presented in this section was published as a detailed study in a preprint [1] and in a conference proceedings paper [2].

V. D. Shemelin, S. A. Belomestnykh, *Multipactor in Accelerating Cavities*, Particle Acceleration and Detection, https://doi.org/10.1007/978-3-030-48198-8_7

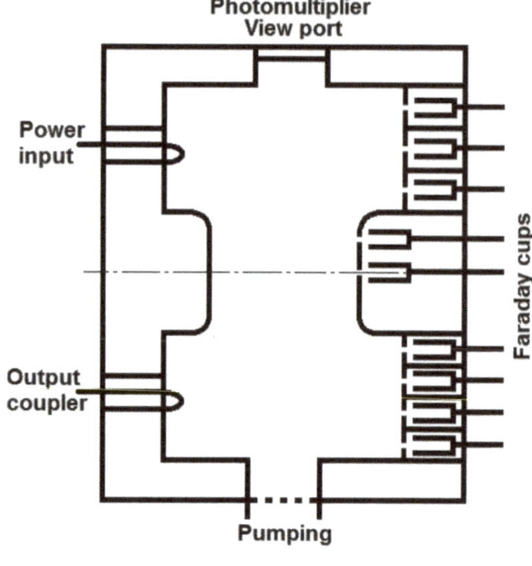

Fig. 7.1 A simplified cross-sectional view of the experimental setup

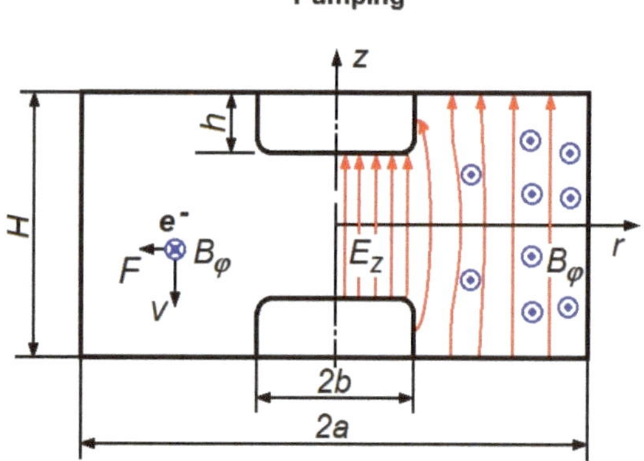

Fig. 7.2 Geometry of the cavity used in the experiment [1] showing the main dimensions as well as electric and magnetic fields and direction of the magnetic component of the Lorentz force

the effect of the ring here and will limit ourselves to a short description of the results needed for evaluation of magnetic field influence on discharge in a flat gap.

Initiation of multipactor was registered by appearance of reflection from the input coupler, glow in the cavity, and distortion of the pulse shape taken from the output loop. The glow in the cavity was observed through a Plexiglas window with a photomultiplier. On the end wall of the cavity, at different distances from the axis, there were nine holes, each 5 mm in diameter. Some incident electrons were trapped by the Faraday cups placed between the copper wall of the cavity and the stainless steel wall of the vacuum volume. The cups were coated with graphite (aquadag) to prevent excessive secondary emission. A bias fed to the Faraday cups was used for measuring the energy of the multipacting electrons.

Multipactor was observed in the central area of the cavity at voltages on axis U_{res} from 7 to 50 kV with two maxima, at 12 and 25 kV. Using the theory presented in Chap. 2, one can estimate that discharge of higher orders, with $n = 4$ and higher, can occur in this area (see Fig. 7.3). The maxima can correspond to the 5th and

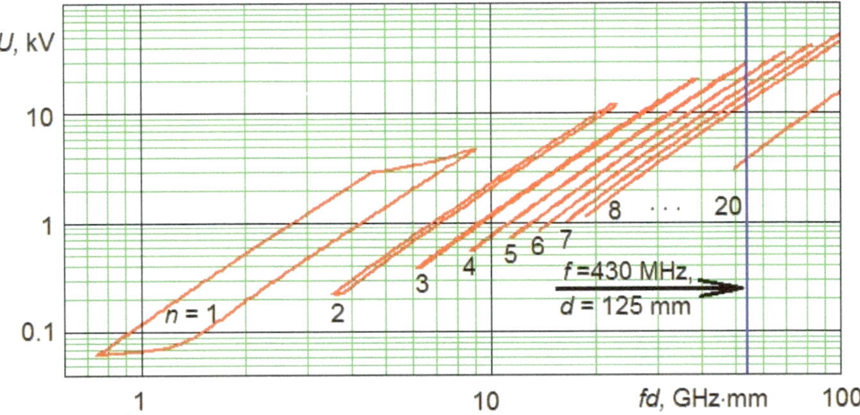

Fig. 7.3 Map of multipacting zones in a flat gap. fd for the gap between reentrant stubs of the test cavity is indicated by a vertical blue line. Zones 9 … 19 are omitted for clarity of the picture. It is assumed that the starting energy of electrons $E_s = 4$ eV and SEY is greater than unity between 50 eV (lower boundary) and 2500 eV (upper boundary of the zones)

6th zones. We cannot expect an exact agreement with the theory and a separation of zones because of the possibility of a generalized phase stability between zones (described in Chap. 3) and because of non-uniformity of the electric field on the reentrant stub: the field at the outer radius is approximately 50% higher than on axis. Nevertheless, the experimental and theoretical boundaries are close. At greater incoming power levels, the discharge shifts to higher radii.

The energy spectrum of incident electrons at $U_{res} = 8$ kV measured on the central Faraday cup falls in the range of 0–600 eV. These values correspond to a higher order multipactor in the central area with $n = 6$ to 8.

It is interesting to note that even the outermost Faraday cup outside the accelerating gap registered a signal at voltages U_{res} from 8 to 12 kV, from 18 to 30 kV, and then from 45 to the maximal achieved voltage of 100 kV. The maximal energy of electrons on this Faraday cup was never higher than 20 eV. As the SEY at such low energies cannot be higher than 1, it was supposed that these electrons appeared as secondary electrons in the inner area and then drifted to the outer radius of the cavity.

This drift can be attributed to the curvature of the electric field force lines toward the cylindrical wall of the cavity (Fig. 7.2). It appears, however, that the RF magnetic field also works for this drift, and this input is significant. In the next section, we analyze this effect in detail.

7.2 Inclusion of Magnetic Field in Equations of Motion[2]

The magnetic field of an RF cavity operating at the fundamental mode increases away from the axis and reaches values that affect multipacting. Conditions can arise under which the particles participating in the discharge are displaced in the radial direction both towards and away from the resonator axis. In this section, we discuss a case when the multipacting electrons cross the flat gap between opposite walls of the cavity but the magnetic field is not negligible anymore as it was supposed in Part I. The case of an elliptical cavity when the particles move near the cavity equator will be discussed in the following chapters.

Here, we consider multipactor in a reentrant cylindrical cavity with the fundamental TM_{010}-like mode of oscillations. We write expressions for the electric and magnetic field in the form

$$E_z = \alpha E_0 \sin \omega t, \tag{7.1}$$

$$B_\varphi = \gamma B_0 \cos \omega t, \tag{7.2}$$

where ω is the oscillation frequency, αE_0 and γB_0 are the amplitudes of the fields, and α and γ specify the dependence on the radius. In particular, α and γ are expressed in terms of Bessel functions for a cylindrical resonator when $h = 0$ (see Fig. 7.2).

We assume that during the flight between walls along z the radial shift is small, so that α and γ vary little during a single flight. We restrict our discussion to the region in which there is no radial electric field; the case when the radial electric field is important is discussed in the next two chapters. The remaining approximations are the same as in the original theory for multipacting in a flat gap (Sect. 2.1).

The equations of motion in the crossed fields (7.1) and (7.2) are of the form

$$m\ddot{z} = e(E_z + \dot{r} B_\varphi), \quad -m\ddot{r} = e\dot{z} B_\varphi, \tag{7.3}$$

where e and m are the charge and mass of an electron, respectively; the dot denotes differentiation with respect to time. The solution of the system (7.3) is [5]

$$z = \frac{1}{\omega} \int y(\theta) d\theta + C_3, \tag{7.4}$$

$$r = \frac{1}{\omega} \int x(\theta) d\theta + D_3, \tag{7.5}$$

[2]Material of Sect. 7.2 was first published in [4].

where

$$
\begin{aligned}
& y = y_0 + C_1 y_1 + C_2 y_2,\ \ x = x_0 + D_1 x_1 + D_2 x_2, \\
& x_1 = y_1 = -\sin(q\sin\theta),\ \ x_2 = y_2 = \cos(q\sin\theta), \\
& x_0 = p\left(x_1 \int x_2 \sin\theta d\theta - x_2 \int x_1 \sin\theta d\theta\right), \\
& y_0 = \frac{p}{q}\left(y_2 \int \frac{y_1}{\cos^2\theta} d\theta - y_1 \int \frac{y_2}{\cos^2\theta} d\theta\right), \\
& p = \frac{e}{m}\frac{\alpha E_0}{\omega},\ \ q = \frac{e}{m}\frac{\gamma B_0}{\omega} = \frac{\gamma\omega_c}{\omega},\ \ \theta = \omega t,
\end{aligned}
$$

where ω_c is the cyclotron frequency.

Usually the magnetic field is small and $q \ll 1$. Correct to the first order, we can assume

$$x_1 = y_1 = -q\sin\theta,\ \ x_2 = y_2 = 1.$$

One can show that in this case (7.4) changes to (2.4) (after appropriate change of variables) – the usual equation of multipactor in a flat gap.

Constants of integration for (7.5) can be found from the conditions

$$\dot{r}(\theta_1) = 0,\ \ \ddot{r}(\theta_1) = -\frac{e}{m} v B = -q\omega^2 \beta_1 d\cos\theta_1.$$

The minus sign in the expression for $\ddot{r}$ is chosen because for the right-hand triple $r,\ \varphi,\ z$ the magnetic component of the Lorentz force is directed to the axis (see Fig. 7.2). After some transformations, we obtain an expression for the radial shift during a single flight across the gap:

$$\Delta r = (pq/4\omega)\Phi_n(\theta_1),$$

where

$$\Phi_n = (2n-1)^2\pi^2 - \left(1 + \frac{2}{k-1}\right) 8\cos^2\theta_1 + \left(1 + \frac{4}{k-1}\right)\sin 2\theta_1. \tag{7.6}$$

n is the order of the multipactor, θ_1 is the equilibrium starting phase of the electrons (the same θ_1 as in Part I), and k is the ratio of the velocity of incident electrons to the initial velocity of the secondary electrons.[3]

[3] Here, we use the obsolete constant-k theory (see Introductory Overview) because it simplifies the analysis but does not influence results significantly.

The value of k cannot be less than 3 because the SEY usually becomes greater than 1 for energies above 40 eV, and the starting energy of secondary electrons is approximately 4 eV or less. Hence, the coefficients for the second and third terms in (7.6) are limited, and the first term defines the sign of Φ_n for $n \geq 2$ as positive. Therefore, the particles that take part in the discharge will drift away from the cavity axis for all orders of multipactor higher than the first.

In the geometry of Fig. 7.2, the curved force lines of the electric field near the protrusion make an additional contribution to this drift from the cavity axis. However, even if the electric field is parallel to the cavity axis, there is a drift of electrons to the minimum of this field, as is shown here and, in a more general case, in Chap. 11.

The experiment reported in [1] and [2] (see the previous section) has shown that the electron current of the discharge appears earlier near the cylindrical wall of the cavity than the voltage in the given region, which is necessary for the existence of discharge, is attained. At the same time, the optimal voltage for the discharge exists at a smaller radius. The current density exceeds that attained in the central part of the gap by two orders of magnitude. The measured energy with which the electrons are incident on the walls amounts to less than 20 eV, i.e., too low for maintaining a self-sustained discharge. The voltages that were attained in the experiment correspond to no less than a third-order multipactor in the large gap. These facts confirm the existence of a radial shift of electrons in a resonant RF discharge caused by the intrinsic magnetic field of the resonator.

References

1. I.K. Sedlyarov, V.D. Shemelin, Resonant high frequency discharge in a reentrant cavity [in Russian]. Preprint INP9-73, Institute of Nuclear Physics, Novosibirsk (1973)
2. V.G. Veshcherevich, I.K. Sedlyarov, V.D. Shemelin, On suppression of a secondary-electron RF discharge in a reentrant cavity of the storage ring VEPP-3 [in Russian]. *Problems of Nuclear Science and Technology. Linear Accelerators*, vol. 1(2), Kharkov (1976)
3. The VEPP-3 electron-positron storage ring. (Budker Institute of Nuclear Physics.) http://v4.inp.nsk.su/vepp3/index.en.html. Accessed 7 Nov 2019
4. V.D. Shemelin, Effect of the intrinsic magnetic field of a volume resonator on the secondary electron discharge in it. Zh. Prikl. Mekh. Tekh. Fiz. **5**, 13 (1981). See also: 0021–8944/81/2205 – 0606 @ 1982 Plenum Publishing Corporation
5. E. Kamke, *Differentialgleichungen Reeller Funktionen*, Band 1 (Akademische Verlagsgesellschaft, Leipzig, 1962)

Chapter 8
Multipactor Near the Cavity Equator

8.1 Introduction

Multipactorin accelerating RF cavities remains a phenomenon that is difficult to predict quantitatively by theory. Especially sensitive to multipacting are superconducting cavities when their main advantage, high quality factor, is spoiled by parasitic electron loading. Early superconducting RF (SRF) cavities suffered from multipacting, which was the dominant limitation of their performance [3]. The invention of round-wall cavities (spherical [4] and elliptical [5]) alleviated the problem, allowing for a significant cavity performance improvement. However, even with the rounded shape of the equator as, for example, in the TESLA cavities [6], multipactor appears sometimes and requires certain time for processing.

8.2 Multipactor in Elliptical Cavities

In 1984, W. Weingarten [7] discovered that the two-sided electron multipactor at the cavity equator is responsible for electron loading in spherically shaped cavities. In his simulations, he also found that the impact energy of primary electrons depends linearly on the starting energy, and for 2 eV, it amounts 30 eV. He remarked that "It is astonishing, that such a relatively low impact energy yields a secondary emission coefficient high enough to give multiplication." Other simulations have found existence of this multipactor in the elliptical TESLA cavity with the results in good agreement with experimental data [8]. Our computer simulations also show that resonant motions of electrons are possible in the TESLA cavity, and that particles have low impact energy (32 eV) at which the SEY should be less than

This chapter is based on materials first published in [1] and [2].

V. D. Shemelin, S. A. Belomestnykh, *Multipactor in Accelerating Cavities*, Particle Acceleration and Detection, https://doi.org/10.1007/978-3-030-48198-8_8

unity. Nevertheless, RF processing is required to eliminate multipacting in this area. The most likely explanation is that niobium – the material superconducting cavities are made of – has a high SEY after wet treatment (see Fig. 8.15) and that cavities are typically not subjected to baking at high enough temperature or to gas discharge cleaning.

Another effect that should be considered is the fact that electrons fall on the surface at an angle not necessarily normal to it, and SEY grows as the angle deviates from normal, see discussion in Sect. 8.7.

Though computer simulations have become the predominant tool for predicting multipactor barriers in RF cavities, they cannot provide a clear guidance on how to change the cavity shape to obviate multipacting. Thus, the search for cavity shapes less prone to multipacting using computer programs is tedious and time consuming.

On the other hand, the results of simulations can be helpful for further analysis of motion and prompt a possible way of better understanding the phenomenon. For example, let us consider the trajectory obtained in multipacting calculations for a reentrant SRF cavity [9], as shown in Fig. 8.1. The calculations were performed with MultiPac [10]. One can see that the dimensions of the trajectory are very small compared with the cavity dimensions. It is clear that the cavity magnetic field has as significant an influence on the motion as the electric field. At the same time, one can assume that the amplitude of the magnetic field is nearly constant within such a small change of coordinates. These features will be used in the derivation of the equations of motion.

Of course, integration of equations of motion is performed in any simulation. However, these are the most general equations taking into account fields in the whole cavity. In this chapter, we try to simplify the electromagnetic field description near the equator and express this field in terms of geometric parameters. The analytical

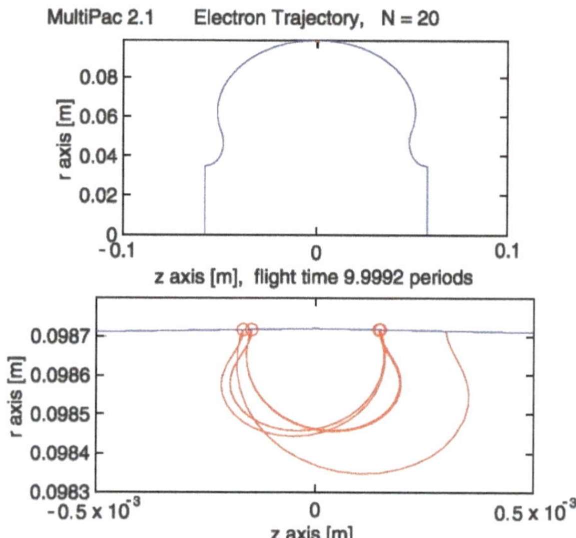

Fig. 8.1 Resonant electron trajectory at a peak surface field level of 37 MV/m

presentation of the equations of motion can give us figures of merit that determine the phenomenon. For example, for multipacting in a flat gap, the value of fd, i.e., the frequency of oscillations times the gap width (or ωd, where $\omega = 2\pi f$), defines the limits of normalized voltage $\xi = eU/m\omega^2 d^2$ across the gap where the discharge takes place (see Chap. 2 or [11]). It appears that in the case of multipacting near the cavity equator, there are also two figures of merit. The first of them is a geometrical parameter depending on the curvature of the surface near the equator and some other geometrical details but not on the field amplitudes: $p = (dE_n/dx)/\omega B_0$, where E_n is the normal to the surface electric field in the area of multipacting, B_0 is the magnetic field in this area, and the derivative is taken along the surface transversely to the magnetic field (Fig. 8.2). The second figure of merit is a parameter proportional to the intensity of fields, or the magnetic parameter $M = eB_0/m\omega$.

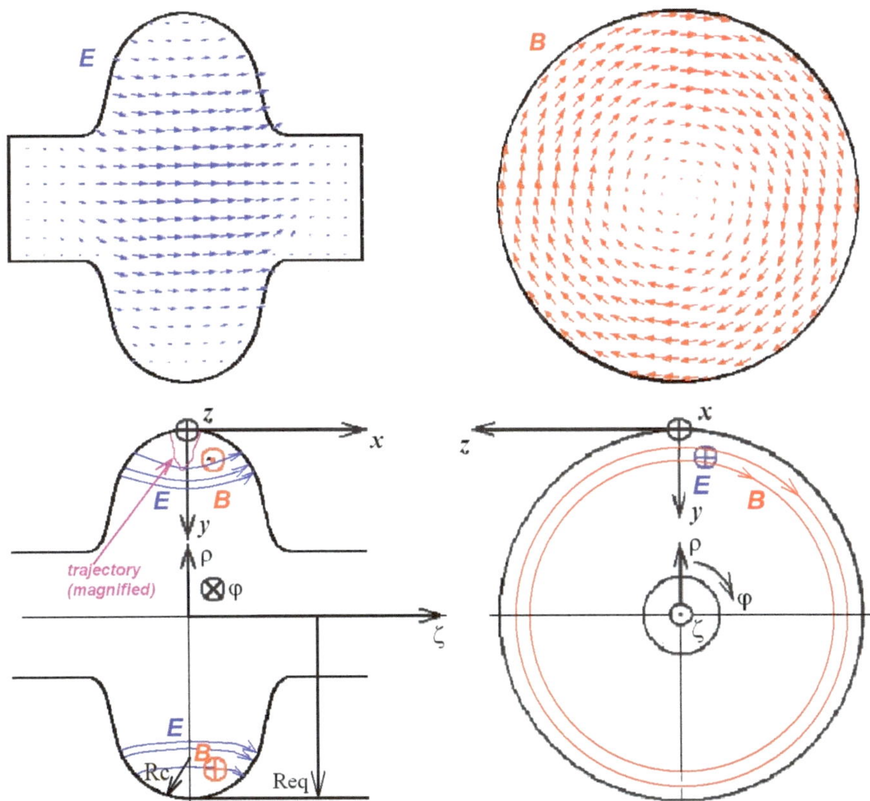

Fig. 8.2 (Top) Electric and magnetic fields in an accelerating cavity. (Bottom) Coordinate systems used for description of multipacting near the equator

8.3 Fields Near Equator

Let us consider the elliptical accelerating cavity shown in Fig. 8.2. The outline of the cavity is formed by two elliptic arcs – one near the equator and another near the beam pipe. There can be a straight segment between these arcs. The top two pictures show the electric (E) and magnetic (B) fields in the cavity.

To analyze the electric and magnetic fields in the immediate vicinity of the equator, we assume that they depend mainly on the radius of curvature R_c near the equator and less on the lengths of the ellipses axes. The equatorial radius of the cavity is designated as R_{eq}. Using the cylindrical system of coordinates $\rho,\ \varphi,\ \zeta$, we can express the fields near the equator in the form of series up to the third order of increments:

$$\begin{aligned} E_\zeta &= \omega B_0 \Big[\alpha_0(R_{eq}-\rho) + \alpha_1(R_{eq}-\rho)^2 + \alpha_2\zeta^2 + \alpha_3(R_{eq}-\rho)^3 + \\ &\qquad\qquad + \alpha_4(R_{eq}-\rho)\zeta^2\Big], \\ E_\rho &= \omega B_0 \Big[\beta_0\zeta + \beta_1(R_{eq}-\rho)\zeta + \beta_2(R_{eq}-\rho)^2\zeta + \beta_3\zeta^3\Big], \\ B_\varphi &= B_0[1 + a(R_{eq}-\rho) + b(R_{eq}-\rho)^2 + h\zeta^2 + q(R_{eq}-\rho)^3 + s(R_{eq}-\rho)\zeta^2]. \end{aligned} \tag{8.1}$$

Not all terms of expansion with a power less than 3 are present because it is taken into account that E_ζ and B_φ are even functions of ζ, and E_ρ is an odd function of this variable.

For calculation of the coefficients $\alpha_0, \ldots, \beta_0, \ldots, a,\ b,\ h, \ldots$ we use Maxwell's equations in the form

$$\vec{\nabla} \times \vec{E} = -\partial \vec{B}/\partial t, \quad \vec{\nabla} \times \vec{H} = \partial \vec{D}/\partial t, \quad \vec{\nabla} \cdot \vec{E} = 0, \tag{8.2}$$

with $\vec{B} = \mu_0 \vec{H}$ and $\vec{D} = \varepsilon_0 \vec{E}$. We will equate coefficients standing before equal terms of the expansion. For the correct comparison of coefficients, it is necessary that the polynomials for fields have equal powers for $R_{eq} - \rho$ and ζ on both sides of the equations. This is why, when substituting fields (8.1) into the first formula from (8.2) we should take only linear and quadratic terms for B_φ, and in calculations with the second formula we should take a more precise expansion of B_φ but only linear and quadratic terms for components of $\vec{E}$.

Substituting (8.1) into (8.2), we obtain the following four groups of equations for coefficients of the expansion.

From $\left[\vec{\nabla} \times \vec{E}\right]_\varphi = -\partial B_\varphi/\partial t$:

$$\begin{aligned} \alpha_0 + \beta_0 &= 1, \\ 2\alpha_1 + \beta_1 &= a, \end{aligned}$$

$$3\alpha_3 + \beta_2 = b, \tag{8.3}$$
$$\alpha_4 + 3\beta_3 = h.$$

From $\left[\vec{\nabla} \times \vec{H}\right]_\rho = \partial D_\rho / \partial t$:

$$-2h = k^2\beta_0,$$
$$-2s = k^2\beta_1, \tag{8.4}$$

where $k = \omega/c$, and $c = 1/\sqrt{\mu_0\varepsilon_0}$ is the speed of light.

From $\left[\vec{\nabla} \times \vec{H}\right]_\zeta = \partial D_\zeta / \partial t$:

$$a = \frac{1}{R_{eq}},$$
$$\frac{1}{R_{eq}^2} + \frac{a}{R_{eq}} - 2b = k^2\alpha_0,$$
$$\frac{1}{R_{eq}^3} + \frac{a}{R_{eq}^2} + \frac{b}{R_{eq}} - 3q = k^2\alpha_1,$$
$$\frac{h}{R_{eq}} - s = k^2\alpha_2. \tag{8.5}$$

From $\vec{\nabla} \cdot \vec{E} = 0$:

$$\frac{\beta_0}{R_{eq}} - \beta_1 + 2\alpha_2 = 0,$$
$$\frac{\beta_0}{R_{eq}^2} + \frac{\beta_1}{R_{eq}} - 2\beta_2 + 2\alpha_4 = 0. \tag{8.6}$$

The equation for $\vec{\nabla} \cdot \vec{H}$ is not included in (8.2) because the only non-zero component, H_φ, does not depend on φ and the equation turns into the identity. The equation for $\vec{\nabla} \times \vec{E}_\zeta$ also gives a trivial result.

Now we have 12 equations and 14 unknowns.

One can then use an additional consideration, that the vector $\vec{E}$ is normal to the surface, as shown in Fig. 8.3. This gives us

$$2\beta_0 = \alpha_0 + 2R_c\alpha_2. \tag{8.7}$$

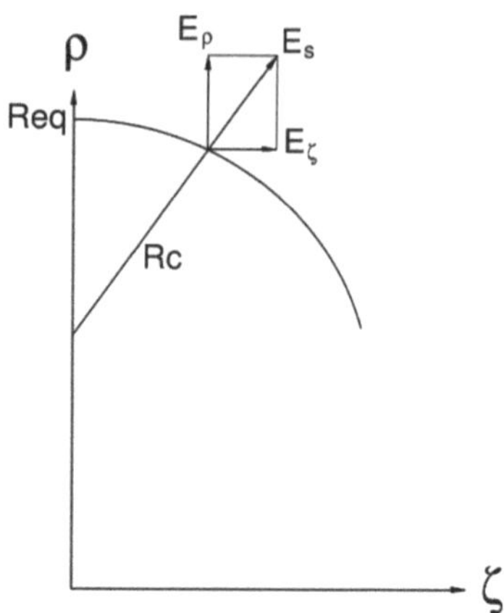

Fig. 8.3 E_s is normal to the surface

Equation (8.7) can be derived from the equality

$$\frac{E_\zeta}{E_\rho} = \frac{\zeta}{\rho - R_{eq} + R_c} \tag{8.8}$$

for the points on the surface:

$$R_{eq} - \rho \approx \zeta^2/2R_c. \tag{8.9}$$

E_ζ and E_ρ should be taken from (8.1), and the lowest power of ζ is used in (8.8).

As will be shown in the next section, fields near the equator do not depend on the local geometry only. Therefore, we are forced to introduce some "external" coefficient. Let us analyze the field B_φ along the cavity surface. This field is almost constant. Let us take into account the quadratic term in

$$B_\varphi = B_0(1 - \nu\zeta^2/2R_c^2),$$

where ζ follows the profile line (8.9) so that $\nu = 1$ for a spherical cavity ($B_\varphi \sim \sin\theta, \quad \theta \approx \pi/2$, where θ is polar angle in the spherical coordinate system). The factor ν shows the degree of deviation from the constant field value along the cavity profile line. Using this equation together with the third one from (8.1) under the condition $(\rho - R_{eq} + R_c)^2 + \zeta^2 = R_c^2$, we obtain the following for the points on the profile line:

$$a + 2R_c h + \nu/R_c = 0. \tag{8.10}$$

Now we have 14 equations. Unfortunately, the last equation in (8.5) is a linear combination of three others and not all the unknowns can be determined. The coefficients for the third powers in the expansions (8.1) for E_ζ and E_ρ, i.e., α_3, α_4, β_2, and β_3, appear undetermined. Therefore, we are forced to limit the accuracy of the expansion for the electric field to the second power of increments and to equate these coefficient to 0. We now rewrite (8.1) in a shorter form:

$$\begin{aligned}
E_\zeta &= \omega B_0 \left[\alpha_0(R_{eq}-\rho)+\alpha_1(R_{eq}-\rho)^2+\alpha_2\zeta^2\right],\\
E_\rho &= \omega B_0 \left[\beta_0\zeta+\beta_1(R_{eq}-\rho)\zeta\right],\\
B_\varphi &= B_0[1+a(R_{eq}-\rho)+b(R_{eq}-\rho)^2+h\zeta^2+\\
&\qquad +q(R_{eq}-\rho)^3+s(R_{eq}-\rho)\zeta^2].
\end{aligned} \tag{8.11}$$

Let us write the chain of equations to determine the remaining coefficients so that each next coefficient can be calculated using the previous ones (explicit expressions are too cumbersome):

$$\begin{aligned}
a &= 1/R_{eq},\\
h &= -a/2R_c - \nu/2R_c^2,\\
\beta_0 &= -2h/k^2,\\
\alpha_0 &= 1-\beta_0,\\
\alpha_2 &= (2\beta_0-\alpha_0)/2R_c,\\
\beta_1 &= 2\alpha_2+\beta_0/R_{eq},\\
s &= -k^2\beta_1/2,\\
b &= 1/R_{eq}^2 - k^2\alpha_0/2,\\
\alpha_1 &= (a-\beta_1)/2,\\
q &= (2/R_{eq}^3+b/R_{eq}-k^2\alpha_1)/3.
\end{aligned} \tag{8.12}$$

The values of α_0 and β_0 that will be required further can also be taken directly from the field calculations.

The validity of the described approach was checked in [2] on two cavity shapes: the TESLA cavity [6] inner cell and a spherical cavity (radius 100 mm, f = 1309 MHz). The fundamental mode fields in both cavities were calculated using SLANS [12]. In addition, fields in the spherical cavity were calculated analytically using expressions from [13]. Then solutions for both geometries were found using the proposed expansion in the area with $|\zeta| \le 5$ mm and $y = R_{eq} - \rho \le 2$ mm. For better accuracy, a 100 × 100 mesh was used in computer simulations.

The differences between the solutions for the TESLA cell are as follows (Figs. 8.4 and 8.5):

$$\delta_E = \frac{|\vec{E}_{slans} - \vec{E}_{\zeta\rho}|}{|\vec{E}_{slans}|} < 1\%,$$

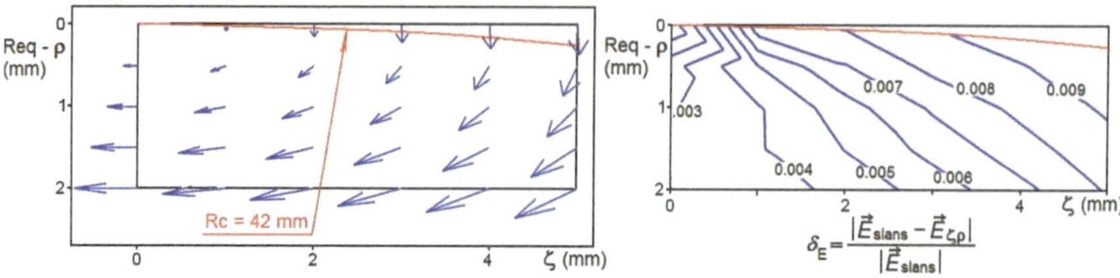

Fig. 8.4 Left: Electric field in the TESLA cavity near the equator. Vectors outside the cavity volume are a formal extrapolation with formulas from (8.11). Right: Difference between results calculated with SLANS and with formulas from (8.11)

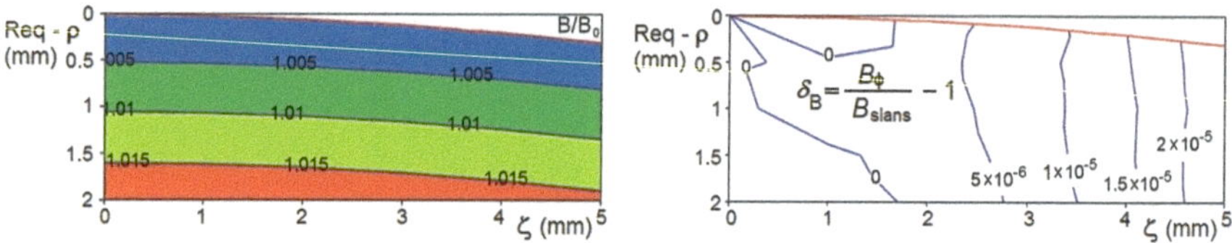

Fig. 8.5 Left: Magnetic field in the TESLA cavity near the equator. Right: Difference between results calculated with SLANS and with a formula from (8.11)

$$\delta_B = \frac{B_\varphi}{B_{slans}} - 1 < 0.0025\%,$$

where $\vec{E}_{\zeta\rho}$, a vector with two components, E_ζ and E_ρ, and B_φ are taken from (8.11).

The magnetic field differs from a constant by approximately 1% for $y \leq 1$ mm and less than 2% for $y \leq 2$ mm. If only linear terms are taken for the electric field, the error is less than 1% for the area $y \leq 1$ mm, $|\zeta| < 2$ mm.

The differences – between the exact analytical solution and SLANS in the same area $|\zeta| \leq 5$ mm, $y \leq 2$ mm – for the spherical cavity are $<0.3\%$ for $\vec{E}$ and $<0.0005\%$ for B_φ. The accuracy of formula (8.11) – the difference from the exact solution for the magnetic field – is even one order of magnitude better in this area.

In the following sections, it will be sufficient to use only linear terms of expansion for the electric field and the magnetic field will be considered constant. However, the results of the expansion can be used to check the accuracy of the field calculations by codes and, possibly, in other applications.

8.4 Dependence of the Upper Arc Fields on the Lower Arc Geometry

In the construction of fields near the equator, on the upper arc, we assumed that the field behaviour depends mainly on the local geometry. However, this assumption does not hold, as shown in Figs. 8.6 and 8.7.

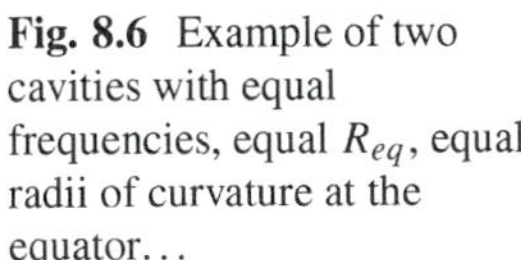

Fig. 8.6 Example of two cavities with equal frequencies, equal R_{eq}, equal radii of curvature at the equator…

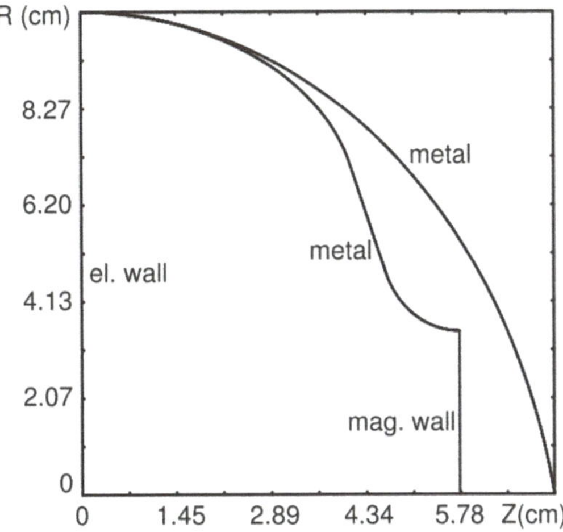

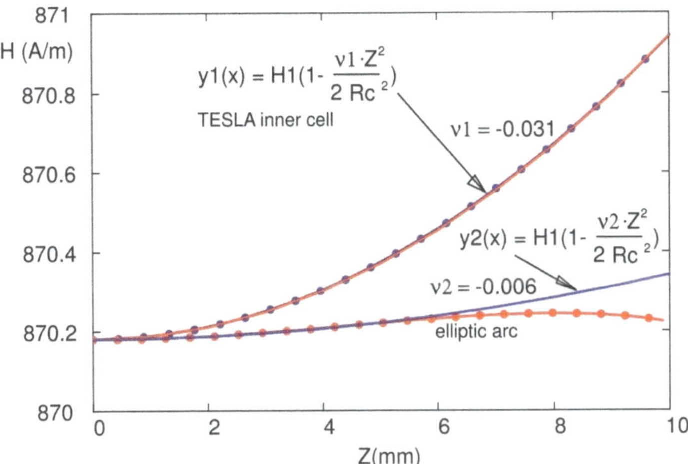

Fig. 8.7 …but with different behavior of fields near the equator

One of the cavities in Fig. 8.6 has the shape of the inner TESLA cell [6] with the magnetic wall at the iris, and another one is described by an elliptic arc with half-axes $A = 73.856$ mm (along the Z-axis) and $B = 129.874$ mm. Both cavities have the fundamental mode frequency of 1300 MHz, equatorial radius $R_{eq} = 103.354$ mm, and curvature radius $R_c = 42$ mm.

The magnetic field on the initial part of the surface is shown in Fig. 8.7. Also shown are the parabolic approximations of these fields in the form

$$y(Z) = H_1 \cdot [1 - \nu \cdot Z^2/(2R_c^2)].$$

Normalization to $2R_c^2$ is used for comparison with the spherical cavity, which has $\nu = 1$.

We can conclude that the behaviour of fields on the upper arc depends not only on geometry of this arc but also on the geometry of the iris area. In particular, it is impossible to make a flat field ($\nu = 0$) on the upper arc by changing this arc only.

8.5 Equations of Motion

The system of coordinates, fields in the equatorial area, and a rough sketch of an electron trajectory in the cavity are shown in Fig. 8.2. Let us present the magnetic and electric fields in this region in the form

$$\begin{aligned} B_z &= -B_0 \cos\theta, \\ E_x &= \alpha \cdot y \cdot \sin\theta, \\ E_y &= -\beta \cdot x \cdot \sin\theta. \end{aligned} \tag{8.13}$$

where $\theta = \omega t$, and α and β are coefficients of proportionality. We will use $\alpha = \alpha_0 \omega B_0$ and $\beta = \beta_0 \omega B_0$, taking α_0 and β_0 from (8.12). Addition of higher order terms of the expansions has an insignificant effect on the results. However, the coefficient h describing the dependence of the magnetic field on ζ^2 should be calculated because it defines values of α_0 and β_0.

Now we can write the equations of motion considering these three components of fields:

$$\begin{aligned} e\,(E_x + \dot{y} B_z) &= m\ddot{x}, \\ e\left(E_y - \dot{x} B_z\right) &= m\ddot{y}. \end{aligned} \tag{8.14}$$

Electron charge is taken as positive for simplicity of writing. Replacing derivatives with respect to time (designated by dots) by derivatives with respect to the phase angle θ (designated by primes), so that

$$\dot{x} = \omega x', \ \ddot{x} = \omega^2 x'', \text{ and so on,}$$

one can obtain the equations of motion in normalized form:

$$\begin{aligned} x'' &= M[(1-p) y \sin\theta - y' \cos\theta], \\ y'' &= M(-px \sin\theta + x' \cos\theta). \end{aligned} \tag{8.15}$$

Here, $M = eB_0/m\omega$ is the magnetic parameter of the motion, and $p = \beta/(\alpha + \beta) = \beta_0$ is the geometrical parameter – introduced in Sect. 8.3 – that depends on the curvature of the surface near the equator and some inner geometrical details (value of ν) but not on the field amplitudes: $p = (dE_n/dx)/\omega B_0$, where E_n is the surface electric field in the area of multipacting. From (8.12), one can obtain

$$p = (c^2/\omega^2) \cdot (a/R_c + v/R_c^2). \tag{8.16}$$

The set of equations (8.15) was solved in MathCAD [14] by the Runge-Kutta method.

A handy way of fast calculation of the parameter p is shown in Sect. 10.4 (see there Fig. 10.5). The electric field at the two points inside the cavity can be obtained with any 2D program (e.g., SuperLANS [12]).

8.6 Condition of Stability

In the case when multipacting occurs near the cavity equator, the starting electron should be described both by the phase of the field and the distance from the equator because the electric field E_y changes with distance x (see (8.13)). If the electron has an initial phase and position different from the equilibrium ones, after the flight to the next impinge onto the surface, its phase and position will change:

$$\begin{aligned} \Delta\theta_2 &= (\partial\theta_2/\partial\theta_1)\Delta\theta_1 + (\partial\theta_2/\partial x_1)\Delta x_1, \\ \Delta x_2 &= (\partial x_2/\partial\theta_1)\Delta\theta_1 + (\partial x_2/\partial x_1)\Delta x_1, \end{aligned} \tag{8.17}$$

where $\Delta\theta_1$ and Δx_1 are deviations from the equilibrium phase and position, respectively, at the start point, and $\Delta\theta_2$ and Δx_2 are deviations after the flight. The derivatives in (8.17) are taken for the equilibrium situation when the time of flight is equal to an integer odd number of half-periods:

$$\theta_2 - \theta_1 = (2n - 1)\pi.$$

The system (8.17) can be written as a matrix product:

$$\begin{pmatrix} \Delta\theta_2 \\ \Delta x_2 \end{pmatrix} = \begin{pmatrix} a & b \\ c & d \end{pmatrix} \begin{pmatrix} \Delta\theta_1 \\ \Delta x_1 \end{pmatrix} \equiv A \begin{pmatrix} \Delta\theta_1 \\ \Delta x_1 \end{pmatrix}.$$

After the Nth flight, the deviations are

$$\begin{pmatrix} \Delta\theta_{N+1} \\ \Delta x_{N+1} \end{pmatrix} = A^N \begin{pmatrix} \Delta\theta_1 \\ \Delta x_1 \end{pmatrix}.$$

Therefore, for stability, it is necessary that

$$\lim_{N\to\infty} A^N = 0,$$

and this requirement is fulfilled if the characteristic roots of the matrix equation

$$A - \lambda I = 0, \tag{8.18}$$

where I is a unitary matrix, meet the conditions

$$|\lambda_1| < 1, \ |\lambda_2| < 1. \tag{8.19}$$

The derivatives for (8.17) were found by numerically solving the equations of motion (8.15) for different p and M and substituted into equation (8.18).

8.7 Multipactor Maps

The areas where the conditions (8.19) are met are shown in Figs. 8.8 and 8.9 for the first and second order two-point multipactor, respectively. The "boundary of stability" is defined as the locus of points where either $|\lambda_1| = 1, \ |\lambda_2| < 1$ or $|\lambda_1| < 1, \ |\lambda_2| = 1$. Lines of the constant impact energy E_p are plotted as well. The values of the impact energy are calculated for an emission energy of

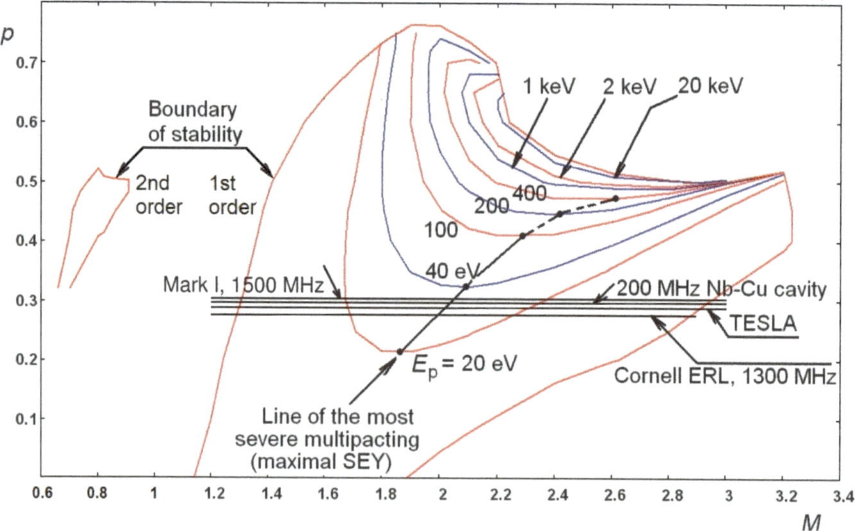

Fig. 8.8 Equatorial crossed-field two-point fist-order and second-order multipacting zones. Emission energy of secondary electrons $E_s = 2\,\text{eV}$. Horizontal lines represent several geometries: Mark I cavity [15], 200 MHz Nb–Cu cavity [16], TESLA cavity [6], and Cornell ERL cavity [17]. E_p is the impact energy (primary electrons). For more details on the second-order zone, see Figs. 8.9 and 8.11

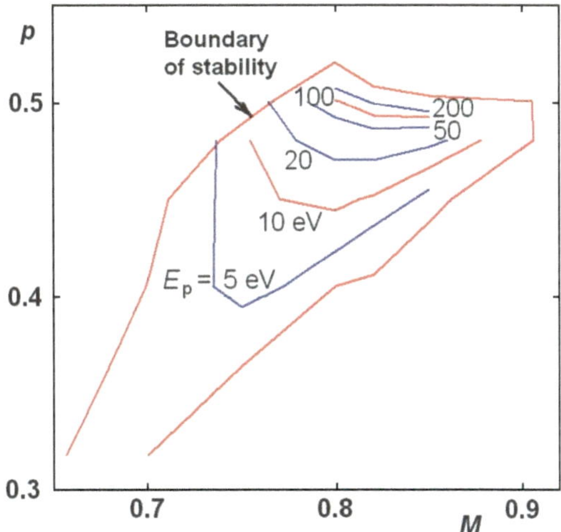

Fig. 8.9 Equatorial cross-field two-point second-order multipacting zone

electrons of 2 eV. The multipacting zone lies within the boundary of stability and is limited by a line where the energy of the impact electrons is high enough to produce SEY > 1. This line should be obtained taking into account the SEY of the cavity wall material and additional considerations described below (the angle of incidence). A particular cavity geometry is represented on the multipactor map by a horizontal line $p = \text{const}$. As we are considering multipactor in elliptical cavities, it is interesting to compare these geometries with an "ultimate" case of the spherical cavity. Analytical solutions for eigenmodes' fields in the spherical cavity are available, e.g., in Ref. [13]. Calculating the electric field at two points, as shown in Fig. 10.5, we find $p = 0.265$. From Fig. 8.8 and Table 8.1 we can see that this is the lowest geometrical parameter among all considered cavities, and hence the spherical cavity would be the least susceptible to multipactor near the equator if such a geometry were practical for acceleration of particles.

A very important role of the emission energy of the secondary electrons should be emphasized: an increase of this energy leads to a proportional increase of the impact energy of the electrons. This fact was obtained from simulations by W. Weingarten [7], who was the first to describe multipactor on the cavity equator. Now we see that this scalability of parameters follows from the homogeneity of equations (8.15): if all the functions – $x(\theta)$, $y(\theta)$, and their derivatives – are multiplied by a constant, the solutions of the equations remain the same. For example, if we assume an emission energy of 4 eV instead of 2 eV as in Figs. 8.8 and 8.9, then we have to double the values assigned to the lines of the constant impact energy. The sizes of trajectories as well as velocities and accelerations at each point on the trajectory increase by a factor of $\sqrt{2}$. The multipacting zone would increase because the line where SEY becomes greater than unity would shift. Thus, one should select a realistic value for the emission energy to estimate multipacting zones correctly. The scalability is not inherent, for example, to multipactor in a flat gap.

An important outcome of this analysis is the fact that the sizes of orbits of electrons participating in multipactor near the cavity equator are proportional to the wave length. They are of the order of 1 mm or less for $f = 1300$ MHz, as shown in Figs. 8.10 and 8.11.

If the frequency changes, the initial normalized velocity of secondary electrons changes also: $x' = \dot{x}/\omega$, see Eqs. (8.14) and (8.15). Because the actual initial velocity does not depend on frequency, the initial normalized velocity is inversely proportional to frequency. To keep solutions of the equation (8.15) the same, the initial distance of the secondary electrons from the center of coordinates (Fig. 8.2) should increase if the frequency decreases, so that the ratio of x' to x remains the same. The actual velocity is a function of RF phase, but not explicitly of the frequency. This means that acceleration is proportional to frequency: for lower frequency actual acceleration is lower but the velocity is the same at the same RF phase because the time for acceleration is increased. Hence fields in the multipactor area are proportional to frequency similar to the acceleration. Following that we can conclude that $M = eB_0/m\omega$ does not depend on frequency. The geometrical parameter p also does not depend on frequency, because it is a ratio of the electric field components at two points near the multipactor area. So, the multipacting maps, Figs. 8.8, 8.9, 8.10, and 8.11, are valid for any frequency.

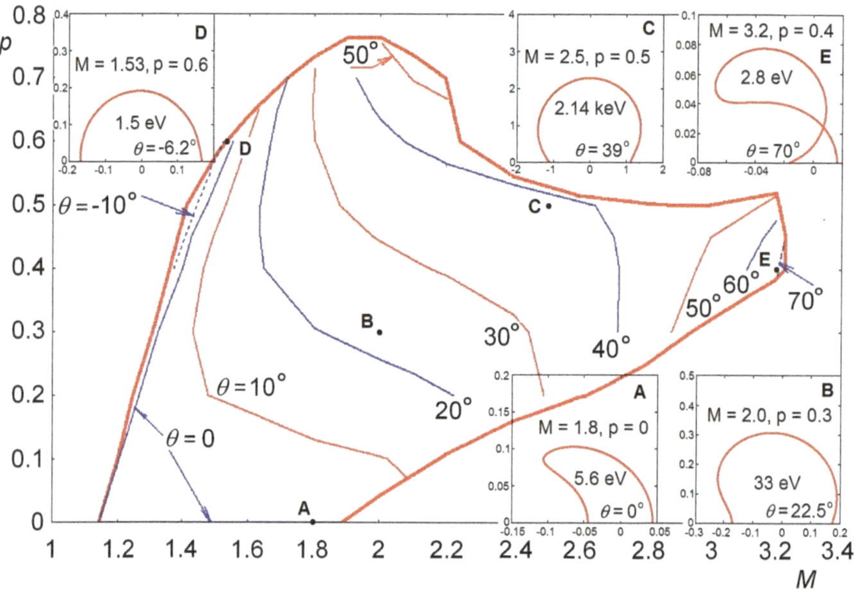

Fig. 8.10 Incidence angle of the impacting electrons in the first zone of the two-point multipactor. Initial velocity vector is normal to the surface. Shapes of trajectories are shown for several points in the area of stability. Dimensions of the trajectories are in millimeters. The trajectories are plotted at different scales but always equal for both axes to show the angle

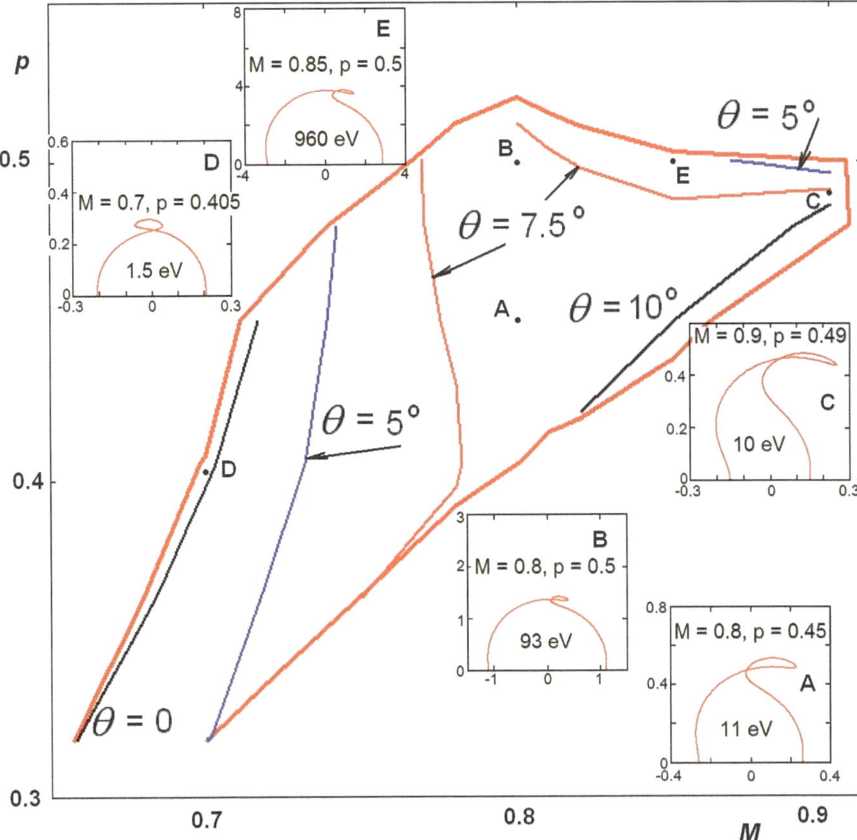

Fig. 8.11 Incidence angle of electrons in the second zone of the two-point multipactor. See explanations in Fig. 8.10

In Sect. 3.2, it was shown that the defining value is the most probable velocity of starting electrons and not the most probable energy, as is usually accepted. The most probable velocity corresponds to a higher value of energy than the most probable energy. This higher energy should be used for calculation of the starting velocity, and the paradox of high SEY at a low energy of primary electrons can be resolved; the starting energy is actually higher than 2 or even 4 eV, and the impact energy is proportionally higher. This value of starting energy can be found from a spectrum of secondary electrons like it was done in Sect. 3.2 for silver.

Parameters M and p are dimensionless. However, the energy on these maps is in electron-volts because we actually calculate the ratio of start and impact velocities and assign a value to the start velocity.

One more important feature of multipacting in crossed RF fields is the angle at which the electrons impinge the surface. Dependence of the SEY on the angle of incidence according to [18] is

$$\delta = \delta_0 e^{\kappa(1-\cos\theta)},$$

where δ_0 is the SEY for the electrons falling normally to the surface, the angle θ is measured from this normal, and the coefficient κ can be found from the data presented in [18]. Though this coefficient is not known for niobium, we can roughly estimate it knowing that it changes from 1.25 for aluminium to 0.75 for molybdenum, and it is close to 1 for copper. If we assume that $\kappa = 1$ for niobium, then we have an increase $\delta/\delta_0 = 1.08$ for the point $M = 2$, $p = 0.3$, where $\theta = 22.5°$. This increase grows fast with the angle, so for $\theta = 60°$ it is 65%. The angle of incidence for different points of the multipacting zones is shown in Fig. 8.10 for the first order and in Fig. 8.11 for the second order of this two-point multipacting. The shapes of trajectories are shown for several points of the area of stable resonant motion. Maximal impact energy of electrons for each value of p is reached on the line connecting the lowest points of the lines of constant E_p (see Fig. 8.8). Due to the increase of the angle with M, as can be seen from the graphs, SEY will also increase, the line of the most severe multipacting will be shifted to higher values of M from these lowest points of the curves of constant energy (Fig. 8.8), and SEY can be higher than 1 even though it is less than 1 for the normal incidence. The incidence angle for the second order multipacting does not vary as widely as for the first order.

8.8 Deviations from the Elliptical Geometry

One can see from Figs. 8.8, 8.9, 8.10, and 8.11 a strong dependence of the multipacting conditions on the geometrical parameter p. As the dimensions of the trajectory are very small, even small deviations from an ideal elliptical shape (compared to the size of the cavity) can change the geometrical conditions – and hence the parameter p – significantly.

The cavity equator is usually a place where two half-cells are welded together (e.g., see Section 6.5 in [3]). A good weld results in a smooth "underbead" at the weld seam. In some cases, especially when weld defects are found, the weld seam is ground down to remove the roughness. In the case when the equatorial region is machined flat, the change of the parameter p is insignificant. The geometry with a flat equator is shown in Fig. 8.12. The value of p does not become zero as could be expected from (8.16) for $R_c = \infty$ in the case of a flat surface, as the size of the flat region is small and the force lines' shapes do not change much. The force lines reflect the ratio between α and β from (8.13) because they have elliptical shapes with the ratio of axis equal to $\sqrt{\beta/\alpha}$, as follows from

$$\frac{E_y}{E_x} = \frac{dy}{dx} = \frac{-\beta x}{\alpha y}.$$

The SLANS calculations for α and β show that $p = \beta/(\alpha + \beta)$ changes from $p = 0.29$ for the original shape to $p = 0.26$ for the width of the flat region $w =$

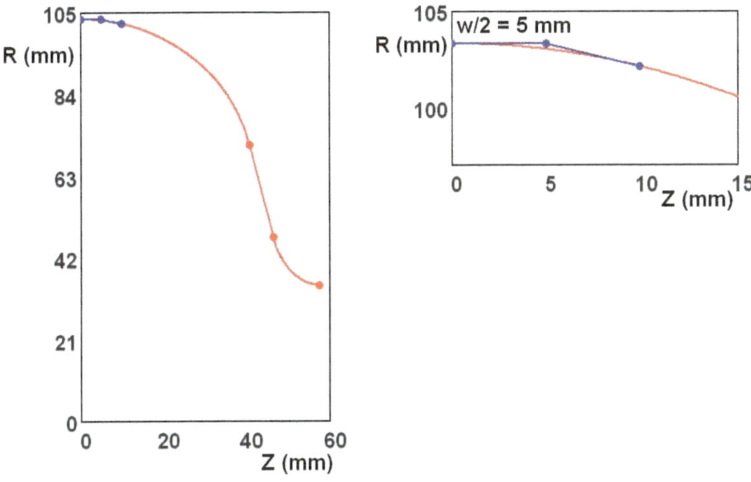

Fig. 8.12 Flat equator in the case of the TESLA cavity (blue line)

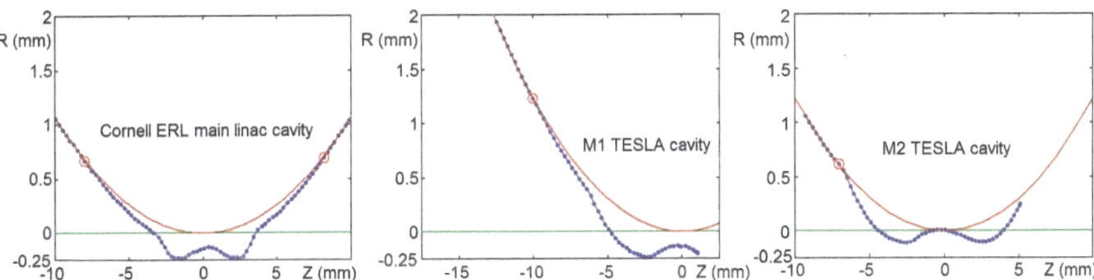

Fig. 8.13 Weld seam profiles for the cavities made by Cornell, Manufacturer 1 (M1), and Manufacturer 2 (M2)

10 mm. Therefore, a small flat region at the equator doesn't change the value of this parameter significantly, and if it changes, it makes it smaller, hampering the emergence of multipacting. Very large flat regions, which can lead to a one-point multipactor [4], will be discussed in the next chapter.

In the case of a protruded bead along the equator, the picture can change substantially. To understand the real shape of the weld seams, measurements of replica molds were made with the help of an ACCURA™ coordinate measuring machine from Carl Zeiss. The replica molds were prepared to study defects on the cavity surface near the equator [19], not specifically for the weld seam study, so not all the seam can be seen in all cases. The replicas were made for the Cornell ERL main linac cavity [20] and for two TESLA cavities fabricated by two companies, which we will refer to as Manufacturer 1 (M1) and Manufacturer 2 (M2). Results of measurements are presented in Fig. 8.13. All the seams are shown at the same scale (note that the scale is approximately 10 times larger in the radial direction) for easier comparison. The seam boundaries – chosen somewhat arbitrarily – are indicated by red circles. Points outside the seam boundaries were used for RMS fitting of the cavity profile to the ideal circular shape, shown in red. The radius of this circle was taken as equal to the radius of curvature at the equator: 48 mm for the ERL cavity and 42 mm for the TESLA cavities.

Comparison of sizes of the weld seams and of the trajectories at these points shows that the seam size (5–10 mm in Z direction) is several times larger than the typical size of the multipacting orbit. The curvature of the surface defining the value of p (8.16) changes within this orbit compared to the design value. Therefore, the seam could have a dramatic effect on the multipacting emergence. However, the SLANS calculation with a shape close to the shape of the TESLA cavity made by Manufacturer 2 (see Fig. 8.13) with the equator seam area approximated by four conjugated circular arcs for the half-cell, gave the value of p close to that of the ideal shape, $p = 0.27$.

Because the central bead of the seam has an insignificant effect on the geometrical parameter p, let us simplify the seam, as shown in Fig. 8.14, and consider two cavities: the 1300 MHz TESLA cavity and a 200 MHz sputtered niobium on copper (Nb/Cu) cavity [16]. In this case, the seam area is described by two elliptical arcs conjugated to one another and to the main circle of the equatorial area. The seam size shouldn't depend too much on the cavity size as both cavities were welded by the same producer, so for the 200 MHz cavity, the same seam ellipses' parameters are used. As noted by R. L. Geng in [21], "The copper cavity [before the Nb film deposition] was electron beam welded from both inside and outside. This practice was sometimes also used for making niobium cavities. The general tendency is that this results in a smaller under-bead because of less heat deposition during each of the two passes of the electron beam. This in turn means that there is a high probability that there is little or no "bump" at center of equator" (at $Z = 0$ in profiles in Fig. 8.13). As shown in Fig. 8.14, the radius of curvature at the lowest point of the seam is greater than the radius of the ideal shape for the TESLA cavity and smaller for the 200 MHz cavity. Calculations of the 200 MHz cavity with a seam give us

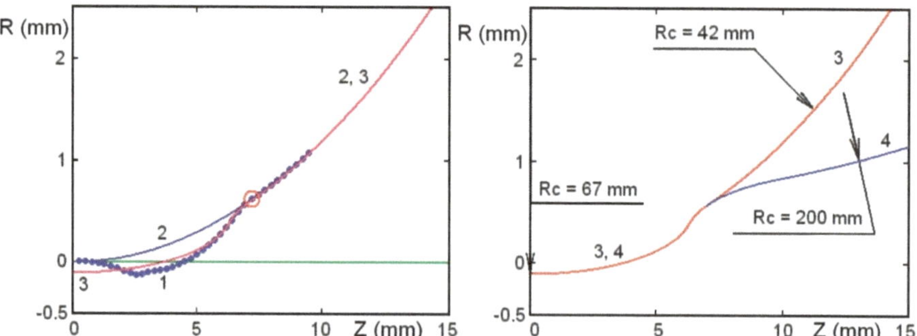

Fig. 8.14 Simplified approximation of the cavity seam profile for the TESLA cavity made by Manufacturer 2 and for the 200 MHz Nb/Cu cavity fabricated at CERN and welded by Manufacturer 2. "1" indicates the measured profile of the TESLA cavity; "2" indicates the ideal contour with a radius of 42 mm; "3" indicates the approximation of the TESLA seam profile by 2 ellipses; and "4" indicates the approximation of the 200 MHz cavity seam profile

the parameter p value of 0.35, instead of 0.296 for the ideal shape. This moves this cavity into the vicinity of the second order multipacting zone and may support the experimental observations described in the next section.

8.9 Comparison with Experiment

Any given cavity geometry has its own value of p. The most severe multipacting (if any) should be expected at the value of M corresponding to the crossing of the line of maximal SEY and the line of the given p. Four examples of cavities with calculated and observed values of multipacting fields are presented in Table 8.1.

The column "E_{acc} calculated" lists the points where the most severe multipactor is expected, except for the Mark I cavity, results for which are explained below in more details.

One can see good agreement between calculated and observed fields. Some difference can be explained by the fact that the discharge starts below the most severe (calculated) point and terminates above it. This is especially true for higher values of p when the energy of primary electrons E_p, and consequently SEY, increases (Fig. 8.8). Higher values of p correspond to higher SEY. The value of SEY = 1 is a threshold. This threshold approximately corresponds to $p = 0.3$ but depends on the surface condition. As can be seen from the table, the value $p = 0.286$ of the TESLA cavity is below this threshold, and although multipactor appears, it can be easily processed. Cavities with higher p have more intense multipactor, and the Cornell cavity, having the smallest p is not prone to multipactor at all.

Two multipacting zones were experimentally observed in the 200 MHz Nb/Cu cavity [16]: one near $E_{acc} = 1$ MV/m and the other in the interval $E_{acc} = 2.8 \ldots 4.0$ MV/m. These values correspond to $M = 0.8$ and $M = 1.7$ to 2.4. From Figs. 8.10 and 8.11, one can conclude that p should be in the range of 0.4 to 0.5 for the first case and 0.2 to 0.55 for the second case.

Table 8.1 Parameters of four cavities and fields of calculated (for the ideal shapes) and observed multipacting. R_{eq} and R_c are in mm, B_0/E_{acc} in mT/(MV/m), and E_{acc} is in MV/m

	R_{eq}	R_c	ν	p	M	B_0/E_{acc}	E_{acc}, calculated	E_{acc}, observed
Nb/Cu cavity	685	200	−.0841	0.296	2.03	4.34	3.3	2.8 – 4
[16], 200 MHz				≈ 0.5 (?)	0.78		1.3	1
TESLA [22, 23], 1300 MHz	103.4	42	−0.0313	0.286	2.0	4.20	22	20, 17– 20
Cornell ERL [17], 1300 MHz	102.9	48.1	0.0068	0.276	2.0	4.20	22	not observed up to 27 MV/m
Mark I [15],	92.02	29.63	−0.0618	0.303	1.44	4.5	17.6	17.6
1500 MHz					2.05		(25.0)	limited by 21 MV/m

As noted in the previous section, calculations with a weld seam shown in Fig. 8.14 gave us $p = 0.35$. Furthermore, as the exact shape of the seam is not known, it is quite possible that the weld seam is even deeper in this cavity than assumed in the figure. This cavity was welded from the inside, contrary to the usual practice of outside welding of the equator of smaller cavities. In addition, further mechanical smoothing was done by locally grinding all the sharp points of the internal surface, as reported in [16].

One more consideration is that Figs. 8.8, 8.9, 8.10, and 8.11 are plotted for the energy of starting electrons $E_s = 2\,\text{eV}$. If this energy is higher, see, for example Figs. 2.3, 3.2, 3.3 and arguments in favor of mean velocity instead of mean energy in Sect. 8.7, the energy E_p in Figs. 8.8 and 8.9 should be proportionally increased. Besides, the incident angle for the first zone (Fig. 8.10) increases the SEY at, e.g., point B by 8%, as stated in Sect. 8.7.

These factors support the possibility of multipactor at point B in the first zone; E_p can be higher than 33 eV, as shown in Fig. 8.10, and in the second zone, the value of p can be less than 0.5 because E_p is actually higher than shown for the point B in Fig. 8.11.

A detailed study of multipacting in single-cell 1.5 GHz SRF cavities was done by J. Knobloch et al. [15]. Results of simulations presented in this work for the cavity Mark I (in which a repetitive multipacting was observed) give the same starting value as can be obtained from Fig. 8.8 ($E_{pk} = 32\,\text{MV/m}$, or $E_{acc} = 17.6\,\text{MV/m}$, $M = 1.45$) if we accept the same values of E_s (3 eV) and SEY = 1 at $E_p = 20\,\text{eV}$ as in the paper [15]. This cavity was prepared using the wet treatment technique, which results in large SEY over the entire range of impact energies (see Fig. 8.15). Two

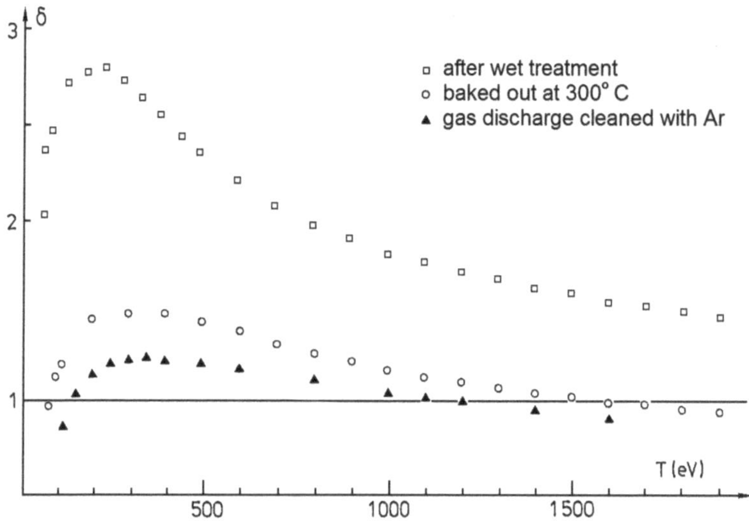

Fig. 8.15 SEY of a niobium surface after different surface treatments as a function of the energy of the impacting electrons [26]

factors, high SEY and big value of p, make this cavity more prone to multipactor than the other three. Argon or helium discharge cleaning (the latter used on the Nb/Cu cavity) or a bakeout (used for Cornell ERL and TESLA cavities), which both remove surface adsorbates, are very effective in reducing SEY.

Apparently, the most severe multipacting was not reached in the Mark I cavity because the maximal value of E_{pk} was limited to 38 MV/m ($E_{acc} = 21$ MV/m). The authors refer to field emission and related thermal breakdown as effects that prevented reaching higher fields. However, it was expected "that multipacting would have been active up to about 42.5 MV/m if the cavity had not been limited by thermal breakdown". According to Fig. 8.8, multipacting in this geometry ($p = 0.303$) would exist at even higher fields having the maximum at $M = 2.05$ corresponding to $E_{pk} = 45.5$ MV/m.

Multipacting in this experiment was observed "at several distinct areas in the equator region" though it constantly shifted "to different places of higher SEY around the equator" after desorption of surface adsorbates. We can suppose that not only the quality of the surface but also the shape deviations are potential causes of multipacting at these *distinct* areas.

Because the phenomenon is very sensitive to the value of E_s, it is desirable to have more reliable information about the emission energy of electrons for niobium.

Processing of the multipactor in a cavity may consist not only in lowering the SEY but also changing the starting energy of the secondary electrons because removing the adsorbates changes the work function of electrons, and hence, their emission (starting) energy E_s. As noted above, because of the scalability of this type of multipactor, the energy of the impacting electrons will also proportionally change in this case and lead to a lower SEY. All the literature agrees that multipactor decreases by decreasing SEY. Our theory gives us a hint that the emitting energy also plays a role.

The formula for the magnetic parameter $M = eB_0/m\omega$ can be rewritten in the form

$$B_0\ [\mathrm{mT}] = 35.7M \cdot f\ [\mathrm{GHz}], \tag{8.20}$$

where limits of M depend on the value of p. The first formula of this kind was offered by W. Weingarten [7], the discoverer of the equatorial multipactor:

$$B_0\ [\mathrm{mT}] = 72f\ [\mathrm{GHz}]/(2n-1),$$

where $n = 1, 2, \ldots$ is the order of the two-sided multipactor. Later, several authors [22, 24, 25] offered analogous formulae with different coefficients before f.

Comparing (8.20) with the phenomenological formulae from the cited papers, one can find the corresponding values of the parameter M, as shown in Table 8.2. The first formula by Fabbricatore et al. [24] follows from an assumption that the frequency of the multipacting motion is the cyclotron frequency for the amplitude of magnetic field and is two times higher than the frequency of the field oscillations. In fact, it is based on the static magnetic field. The second formula from [24] is

Table 8.2 Empirical formulae for the multipacting fields and corresponding values of M

Author	Formula	M calculated from (8.20)
Weingarten (1984) [7]	B [mT] = 72 f [GHz]	2.02
Fabbricatore et al. (1995) [24]	B [mT] = 78 f [GHz]	2.18
	B [mT] = 55 f [GHz]	1.54
	B [mT] = 57 f [GHz]	1.60
Saito (2001) [22]	B [mT] = 60 f [GHz]	1.68
Geng (2003) [25]	B [mT] = 5 + 55 f [GHz]	1.54 +0.14/f

a correction of the first one by changing the amplitude of magnetic field to its RMS value, dividing it by $\sqrt{2}$. In the next line, there is a value corresponding to an experiment with a 3 GHz cavity. We can see that this guess and the experiment are in a good agreement but data by other authors have deviations from this formula.

Cavities analyzed by different authors differ in shape, which can explain the difference in the values of the coefficients before f in Table 8.2. R. Geng made an attempt to unite different data involving dependencies on frequency, but apparently the shift from direct proportionality to frequency was due to different shapes of the cavities, not to their frequencies.

Formula (8.20) generalizes the phenomenological formulae because it reflects dependence on geometry, i.e., field parameter M is defined by the geometric parameter p. This formula can be rewritten in a more practical form. For any cavity, the value B_{pk}/E_{acc} is known as one of the most important figures of merit. If B_0 is the magnetic field at the equator and E_0 is the accelerating field, both corresponding to the presence of multipacting, then $B_{pk}/E_{acc} = B_0(1+\varepsilon)/E_0$. The value of ε shows the deviation of the peak field from the field at the equator, and for a thoroughly optimized cavity, it is small. For example, the TESLA cavity has $\varepsilon = 0.005$, and the Cornell ERL cavity has $\varepsilon = 0$. Therefore, this value can be neglected. From the foregoing expression and from (8.20), we can obtain the accelerating field of the most probable multipacting:

$$E_0 = \frac{B_0(1+\varepsilon)}{B_{pk}/E_{acc}} = \frac{35.7 \cdot M \cdot f}{B_{pk}/E_{acc}} \text{ [units are MV/m, GHz, mT].}$$

The value of M for the most probable multipacting depends on the geometric parameter of the cavity and can be taken, see Fig. 8.8 for example, as equal to 2.0 for $p = 0.27$ or 2.1 for $p = 0.33$. The values $B_{pk}/E_{acc} \approx B_0/E_{acc}$ needed for calculation of the accelerating field at multipacting $E_{acc} \approx E_0$ are given in Table 8.1 for the cavity geometries considered earlier.

8.10 Conclusion

In this chapter, we presented consideration of multipacting near the equator of an elliptical cavity. Electric and magnetic fields in the equatorial region were given in the form of Taylor series and compared with theory and simulations, demonstrating a very good accuracy of the expansion. Using these analytical approximations, zones of the first and second order of the two-point discharge were found and the energy and incident angle maps of both zones were presented. The influence of the equatorial weld seam shape was discussed in detail. The predictions were compared with experimental results. An explanation of the existence of the second order zone in the 200 MHz Nb/Cu cavity was offered. This zone was not found in simulations performed by the authors of [16]. Finally, a clear physical meaning was given to the empirical formulae for the multipacting field obtained by several authors.

References

1. V. Shemelin, Multipacting in crossed RF fields near cavity equator, in *Proceedings of EPAC 2004, European Particle Accelerator Conference*, Lucerne, 2004, p. 1075
2. V. Shemelin, Multipactor in crossed RF fields on the cavity equator. Phys. Rev. ST Accel. Beams **16**, 012002 (2013)
3. H. Padamsee, J. Knobloch, T. Hays, *RF Superconductivity for Accelerators* (Wiley, New York, 1998). ISBN 0-471-15432-6
4. U. Klein, D. Proch, Multipacting in superconducting RF structures, in *Proceedings of Conference on Future Possibilities for Electron Accelerators*, Charlottesville, 1979, p. N1
5. P. Kneisel, R. Vincon, J. Halbritter, First results on elliptically shaped cavities. Nucl. Ints. Methods Phys. Res. A **188**, 669 (1981)
6. B. Aune et al., Superconducting TESLA cavities. Phys. Rev. ST Accel. Beams **3**, 092001 (2000)
7. W. Weingarten, Electron loading, in *Proceedings of 2nd Workshop on RF Superconductivity*, Geneva, 1984, p. 551
8. P. Ylä-Oijala, Electron multipacting in TESLA cavities and input couplers. Part. Accel. **63**, 105 (1999)
9. V. Shemelin, H. Padamsee, R.L. Geng, Optimal cells for TESLA accelerating structure. Nucl. Inst. Methods Phys. Res. A **496**, 1 (2003)
10. P. Ylä-Oijala et al., *Multipac 2.1 – Multipacting simulation toolbox with 2D FEM field solver and MATLAB graphical user interface.* User's Manual (Rolf Nevanlinna Institute, Helsinki, 2001)
11. K. Krebs, H. Meerbach, Die Pendelvervielfachung von Sekundärelektronen. Ann. Phys. **15**, 189 (1955)
12. D.G. Myakishev, V.P. Yakovlev, The new possibilities of Superlans code for evaluation of axisymmetric cavities, in *Proceedings of PAC1995, Particle Accelerator Conference and International Conference on High-Energy Accelerators*, Dallas, 1995, p. 2348
13. S. Belomestnykh, Spherical cavity: analytical formulas. Comparison of computer codes. LEPP Report SRF941208-13, Cornell University (1994). https://www.classe.cornell.edu/public/SRF/1994/SRF941208-13/
14. Wikipedia contributors. Mathcad, in *Wikipedia, The Free Encyclopedia.* Retrieved 22:54, 7 Nov 2019, from https://en.wikipedia.org/wiki/Mathcad

15. J. Knobloch, W. Hartung, H. Padamsee, Multipacting in 1.5-GHz superconducting niobium cavities of the CEBAF shape, in *Proceedings of 8th Workshop on RF Superconductivity*, Abano Terme (Padova), 1997, p. 1017
16. R. Geng et al., First RF test at 4.2 K of a 200 MHz superconducting Nb–Cu cavity, in *Proceedings of PAC2003, Particle Accelerator Conference*, Portland, 2003, p. 1309
17. N. Valles, M. Liepe, Seven-cell cavity optimization for Cornell's energy recovery linac, in *Proceedings of 14th Workshop on RF Superconductivity*, Berlin, 2009, p. 538
18. H. Bruining, *Physics and Applications of Secondary Electron Emission* (McGraw-Hill, New York, 1954)
19. M. Ge et al., Routine characterization of 3D profiles of SRF cavity defects using replica techniques. Supercond. Sci. Technol. **24**(3), 035002 (2010)
20. N. Valles et al., The main linac cavity for Cornell's energy recovery linac: cavity design through horizontal cryomodule prototype test. Nucl. Inst. Methods Phys. Res. A **734**, 23 (2014)
21. R. Geng: Private communication
22. K. Saito, Experimental formula of the on-set level of two-point multipacting over the RF frequency range 500 MHz to 1300 MHz, in *Proceedings of 10th Workshop on RF Superconductivity*, Tsukuba, 2001, p. 419
23. K. Twarowski, L. Lilje, D. Reschke, Multipacting in 9-cell TESLA cavities, in *Proceedings of 11th Workshop on RF Superconductivity*, Lübeck/Travemünde, 2003, p. 733
24. P. Fabbricatore et al., Experimental evidence of MP discharges in spherical cavities at 3 GHz, in *Proceedings of 7th Workshop on RF Superconductivity*, Gif sur Yvette, 1995, p. 385
25. R.L. Geng, Multipacting simulations for superconducting cavities and RF coupler waveguides, in *Proceedings of PAC2003, Particle Accelerator Conference*, Portland, 2003, p. 264
26. H. Piel, Superconducting cavities, in S. Turner (ed.) *CERN Accelerator School: Superconductivity in Particle Accelerators* (CERN Publication 89–04, Hamburg, 1988) pp. 149–196

Chapter 9
One-Point Multipactor in Crossed Fields of RF Cavities

9.1 Introduction

As mentioned earlierin the book,two forms of multipacting are distinguished in RF devices: one-point and two-point multipacting. Sometimes these are called one-surface and two-surface multipacting, respectively. One-point multipacting (we will call it MP1 in the following) is characterized as follows: (i) the electrons are emitted from and returned to the same surface and (ii) the time between surface impacts is an integer number of RF periods. In the case of two-point multipacting (MP2), the electrons are oscillating in a cavity in such a manner that they hit the surface – typically of the opposite cavity walls – after an odd number of half periods (1/2, 3/2, 5/2, . . .).

The number of incomplete periods in the case of MP2 or complete periods in the case of MP1 is called the order of multipacting. Between collisions with walls, the electrons are subjected to the forces from both electric and magnetic fields.

An analysis of the 2-point multipactor near the cavity equator was presented in Chap. 8. Now we will concentrate our attention on MP1 in SRF cavities, analyze some existing experimental data, and compare these two kinds of RF discharges.

We will use a cavity geometry from [3] for further considerations because this paper gives a very detailed and precise description of multipactor as well as results of computer simulations. The software for the simulations is described in another paper [4]. It was possibly the first serious computer program for multipactor simulations. It included a Monte Carlo estimation of the number and momentum of secondary electrons and backscattered electrons, production of photons, and photo-electric and Compton effects. One of the conclusions in this paper was the following:

This approach to MP1 was first considered in [1] and then expanded in [2].

V. D. Shemelin, S. A. Belomestnykh, *Multipactor in Accelerating Cavities*, Particle Acceleration and Detection, https://doi.org/10.1007/978-3-030-48198-8_9

the "electron avalanche multiplication effect ... is non-resonant in nature". One can see a contradiction of this statement with the results from [3], where it is shown that "the multipacting is limited to discrete field levels". The question arises: if these field levels are not resonances, why they are discrete? In the following, we will try to answer this question.

Of course, the simulations reveal the necessary conditions for multipacting, but the kind of the multipactor is defined a posteriori, as a result of the simulation. Understanding the conditions of the existence of this or another kind of multipactor can give us a clue to prevent it.

Due to the small sizes of electron trajectories, it is possible to expand fields near the multipactor site into Taylor series and solve the differential equations of motion in the nearest vicinity of this site. Parameters of these equations become the figures of merit of the multipactor, and the ability to control them provides us with ways to prevent multipacting.

9.2 Fields and Equations of Motion in a Known Geometry with One-Point Multipacting

Descriptions of MP1 can be found, for example, in [3] or [5]. The latter paper deals with the rather complicated geometry of a so-called muffin-tin cavity, and the geometry in the former paper is closer to that of contemporary accelerating cavities. However, trajectories of the multipacting electrons illustrated in both papers are very similar. For further considerations we will use the cavity geometry from the first paper as it is closer to an elliptical cavity shape.

Though not all dimensions of the cavity are presented in the paper, we can approximately define them from the given drawing and frequency of 2.8 GHz. This S-band cavity, shown in Fig. 9.1, has a flat equatorial region. The rounding between this region and the tilted side wall has a radius $R = 4.7$ mm. From this figure, we determined – with some uncertainty – the equatorial radius $R_{eq} = 43.6$ mm and wall slope angle $\alpha = 85°$. The flat equatorial region (absent in elliptical cavities) appears responsible for multipacting.

The fields in the cavity were calculated with SuperLANS [6]. This software gives the best accuracy among the 2D programs but even with this program – by reasons explained later – we had to use a very dense mesh (180×100). The mesh is shown in Fig. 9.1. As pointed out in the figure, multipactor occurs near the place where the flat equatorial part merges with the rounded region.

The surface electric field for different values of radius R is shown in Fig. 9.2, and fields in the volume near this place for $R = 4.7$ mm are presented in Fig. 9.3. One can see that in the vicinity of the multipacting spot with the dimensions 2×2 mm (upper estimation of the size of the particle orbit) the radial component of the electric field E_r deviates from the average value by $\pm 16\%$, the magnetic field H is constant with an accuracy of $\pm 2\%$, and the longitudinal electric field E_z is linear with the distance from the surface. The difference in E_r between the equator and this spot at some distance from the symmetry plane consists in the following: E_r is proportional

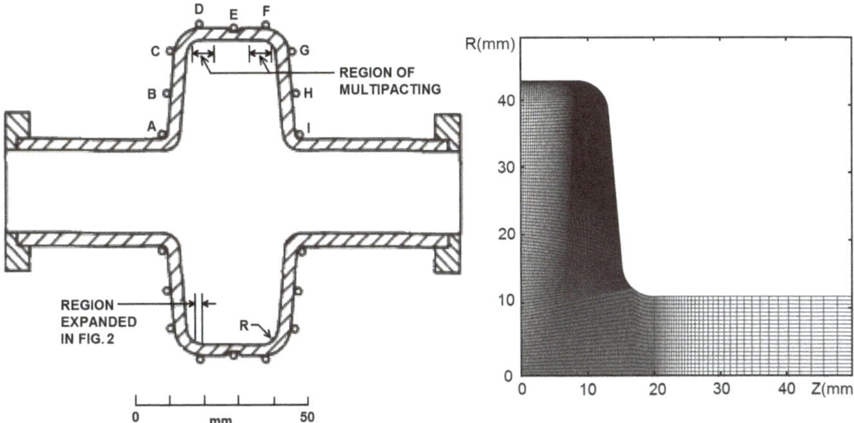

Fig. 9.1 Schematic drawing of the S-band cavity from [3] (left) and geometry with mesh for SLANS calculation (right). (Note: Fig. 2 from [3] is reproduced in Fig. 9.5)

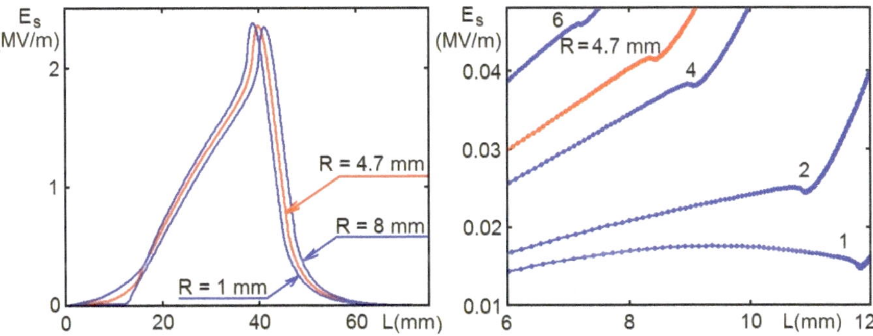

Fig. 9.2 Surface electric field along the profile line of the cavity. L is distance from the equator. The right figure is not a direct expansion of the left one: more curves are to show a shift of the minimum in detail

to the distance from the equator and changes its sign when one passes over the equator, but at some distance, if the size of the spot is smaller than this distance, E_r varies only slightly. This variation of E_r with z can be also accounted for, but we will first simplify our consideration and do the correction later. Figure 9.4 presents the calculation results similar to those in Fig. 9.3 but for $R = 1$ mm.

Up to now, and in Figs. 9.3 and 9.4, we used a "natural" coordinate system – with the origin at the cavity center, axis Z along the axis of rotation and axis R perpendicular to it. For further consideration we will introduce a Cartesian coordinate system.

Placing a new system of coordinates with the origin at the multipactor spot, with an axis x parallel to the axis of the cavity and directed from the equator, with an axis $y = R_{eq} - R$ directed from the surface to the axis of the cavity (R_{eq} is the equatorial radius and R is distance from the cavity axis), and with an axis z supplementing this

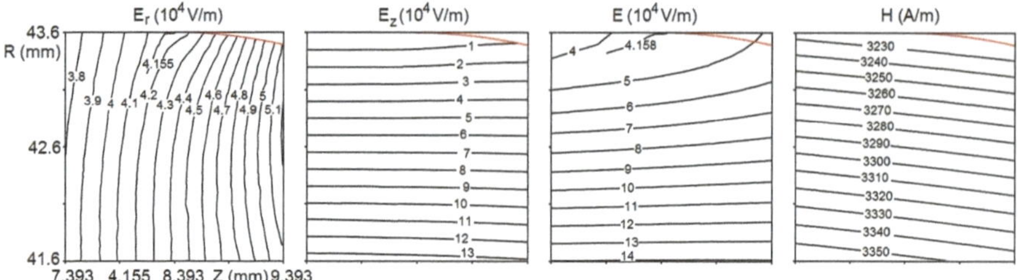

Fig. 9.3 Electric and magnetic fields near the multipacting spot of the cavity from Fig. 9.1. Radius of rounding of the outer cylindrical wall $R = 4.7$ mm. Total energy of field in the cavity is 1 mJ in all the calculations

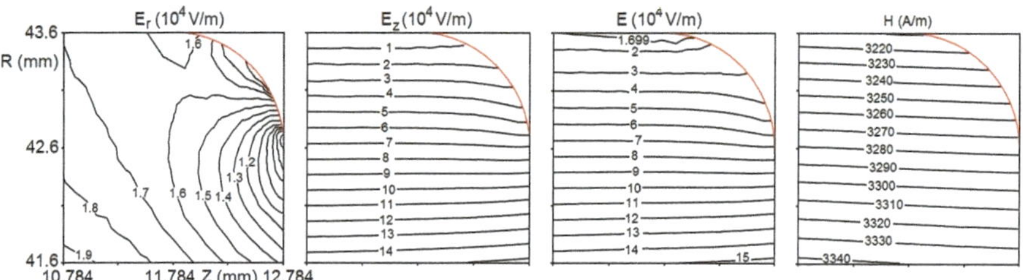

Fig. 9.4 Electric and magnetic fields near the rounded corner with $R = 1$ mm

right-hand system of coordinates, we can present the electric and magnetic fields in the form

$$\begin{aligned} E_x &= -\alpha y \sin\theta, \\ E_y &= E_0 \sin\theta, \\ B_z &= B_0 \cos\theta. \end{aligned} \tag{9.1}$$

where $\theta = \omega t$,

From Maxwell's equation

$$\oint_l \vec{E} d\vec{l} = -\frac{\partial B}{\partial t} \cdot S,$$

for any small contour around the multipacting spot with length l and surface area S, we can find

$$\alpha = \omega B_0. \tag{9.2}$$

Now we can write the equations of motion considering these three components of fields:

$$e\,(E_x + \dot{y}B_z) = m\ddot{x},$$
$$e\left(E_y - \dot{x}B_z\right) = m\ddot{y}. \tag{9.3}$$

These are the same equations as for MP2 (8.14), but the dependencies of fields on coordinates are different. Replacing derivatives with respect to time by derivatives with respect to the phase angle θ so that

$$\dot{x} = \omega x',\ \ \ddot{x} = \omega^2 x'', \text{ and so on,}$$

one can obtain the equations of motion in normalized form:

$$x'' = M(-y\sin\theta + y'\cos\theta),$$
$$y'' = M(q\sin\theta - x'\cos\theta). \tag{9.4}$$

Recall that for MP2 we introduced two dimensionless parameters: magnetic parameter M and geometrical parameter p. For MP1, instead of p, we introduce a different geometrical parameter $q = E_0/\omega B_0$, which depends on the cavity shape but not on the field amplitudes. M is dimensionless and q has a dimension of length, and for the geometry in Fig. 9.1, $q = 0.605$ mm at the initial point of the rounding ($Z = 8.393$, $R = 43.6$ mm, Fig. 9.3). In what follows, the size of the trajectory appears to be of the order of q.

The right part of the first equation in (9.4) is an exact derivative. Therefore, we can integrate this equation and substitute x' into the second one. After this, we have

$$x' = My\cos\theta,$$
$$y'' + M^2 y\cos^2\theta - Mq\sin\theta = 0. \tag{9.5}$$

The first equation in (9.5) means that the x-component of the velocity is always zero when the electron impacts the surface if it had started normally to the surface. Therein lies a difference with the two-point multipacting, where a considerable angle can be between the normal to the surface and the vector of velocity of the impinging electrons, which leads to enhancement of the SEY.

9.3 Comparison of Analytical Calculations with Simulations and Experimental Data

In spite of significant simplifications of the equations of motion and an approximate knowledge of the geometry, the orbits obtained solving equations (9.4) are very similar to the orbits in the cited paper [3] (Fig. 9.5).

Lyneis et al. [3] also reports values of the axial electric fields E_a at which the multipacting intensity was observed to reach a maximum in both experiment

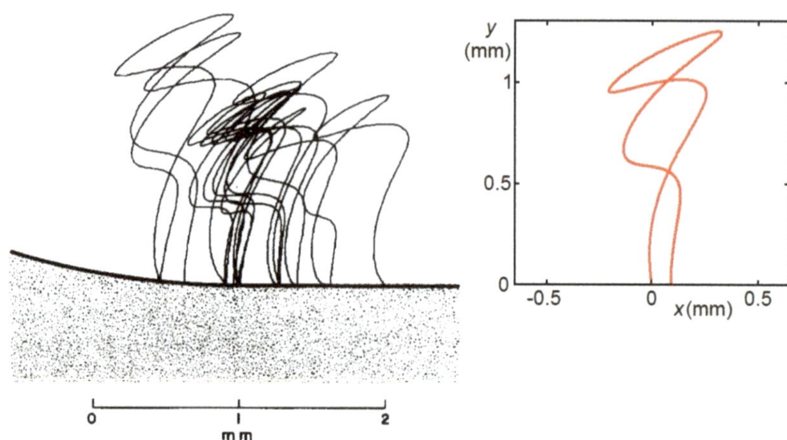

Fig. 9.5 Example of electron orbits from [3] for third-order multipactor and a solution of (9.4) for the S-band cavity geometry from Fig. 9.1

and computer simulations. Of course, in reality there are multipacting bands, not narrow lines, for each order of multipacting. These bands were calculated using (9.4). Boundaries of these bands are defined by the condition of stability introduced in Chap. 2. This condition is satisfied when an electron having an initial phase deviation from equilibrium returns to the surface with the phase deviation reduced. An analogous condition should be used for the coordinate on the surface, in a similar way as it was done in Chap. 8, but it turns out that no space stability for MP1 can be found: the particles always shift to the equator after each flight.

We solved (9.4) for two values of emission energy of secondary electrons: 2 and 4 eV. This means that returning electrons – emitted at a retarding electric field – were taken into account as described in Chap. 2. For the first two orders, the limitation caused by these returning electrons happens earlier than the limitation caused by stability. Boundaries of the multipacting bands found with (9.4) under these conditions are shown in Fig. 9.6.

These boundaries are found under the condition that the secondary emission yield is greater than 1 for all the energies. In the experiment presented in [3], special measures were taken to increase the SEY: the niobium surface was anodized enhancing SEY enough to support multipactor up to the sixth order. If one knows the values of the impact energy (lowest and highest) when SEY becomes 1, the limits of the MP bands can be easily determined by drawing horizontal lines in Fig. 9.6 corresponding to these energies.

In [3], the fields for the first order multipactor are not presented. Results of the simulated and experimentally observed fields at maxima of multipactor from [3] are compared with our analytical results in Fig. 9.7 for orders of multipactor 2 through 6. The calculated maximal impact energies from the simulations are also indicated. One can see good agreement between the analytical results and the experiment. The analytical solutions are much closer to the experimental data than the simulation

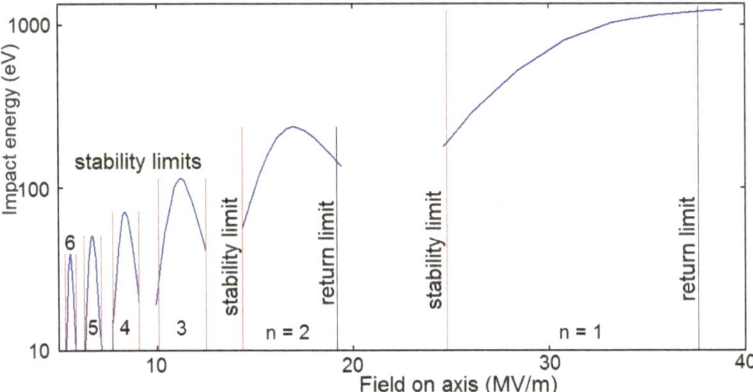

Fig. 9.6 Analytically found boundaries of MP1 bands for the S-band cavity geometry. The emission energy is 4 eV

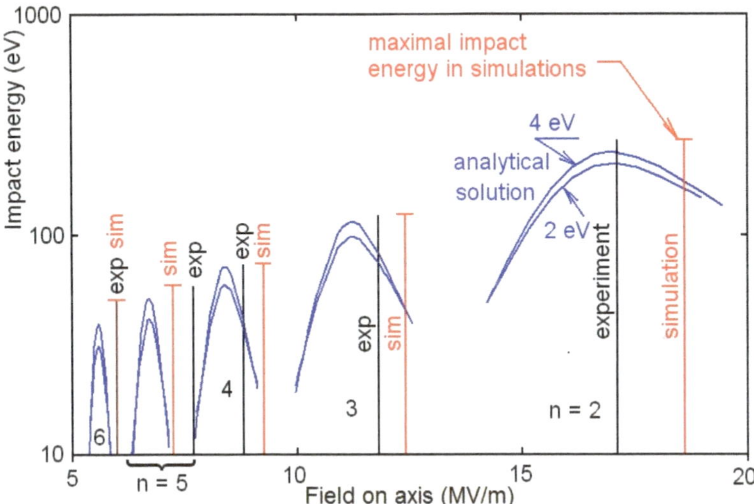

Fig. 9.7 Comparison of results from [3] and analytical results for 2–6 orders of MP1

for the second order and have approximately equal deviation from the experimental data to the simulation for the third and fourth order multipacting bands. Maximal intensity of simulated multipactor is at a higher field than in the experiment for all bands but the fifth, and both simulations and experiment coincide for the sixth band.

The difference between simulations and the analytical solution is regular: analytically found peak fields are always lower than the field found in simulations. A possible explanation for this fact is including backscattering, bremsstrahlung, photo-electric, and Compton effects in the simulations. All of these phenomena are more intense at higher fields and so can shift the peaks to higher values. However, the analytical calculation appears not worse than the simulations if compared with the experiment, which could mean that all these effects have a small influence on the positions of the bands. These effects define the x-ray energy spectrum produced by electrons impacting the cavity walls; the analytical calculation cannot predict this spectrum. Maximal intensity of multipactor apparently occurred at the highest impact energy, and this energy was calculated for two values of the emitted electrons: 2 and 4 eV. We couldn't find a reference for the emission energy used in simulations in the cited papers [3, 4], though it is mentioned that this energy is statistically distributed using a Monte Carlo technique. Comparing maxima of impact energies in Fig. 9.7 for simulations and for analytical solution one can suppose that the mean energy in simulations is close to 4 eV. In the analytical approach, no distribution of energies was used. As can be seen from the results, this has only a slight effect on the widths of the bands.

9.4 Influence of Change of the Surface Electric Field: Multipactor Map

We can take into account changing of the radial component E_r shown in Figs. 9.3 and 9.4 (or E_y in designation of equation (9.1)) along the surface of the cavity. Let us consider the surface component of the electric field E_s as the radial component. This would allow us to include the curved part of the cavity profile line so that x can now be a coordinate along the "straightened" profile line. This is possible if the particle orbit size is much smaller than the radius of curvature. Then, we can change the second equation in (9.1) to

$$E_y = E_s(x) \sin\theta. \tag{9.6}$$

The relationship (9.2) will change to

$$\alpha = \omega B_s - dE_s/dx. \tag{9.7}$$

and the normalized equations of motion (9.4) will take this form:

$$\begin{aligned} x'' &= M[(dq/dx - 1)y \sin\theta + y' \cos\theta], \\ y'' &= M[q(x) \sin\theta - x' \cos\theta]. \end{aligned} \tag{9.8}$$

Now, $q(x) = E_s(x)/\omega B_s(x)$. The behavior of $q(x)$ and dq/dx along the profile line for the S-band cavity geometry (Fig. 9.1) is shown in Fig. 9.8 for $R = 4.7$ mm and $R = 0.38$ mm. The origin of the coordinate system is placed at the equator.

Practically, $q(x)$ is proportional to the surface field $E_s(x)$ in Fig. 9.2 because $B_s(x)$ only weakly changes along the profile line. dq/dx at the initial point of the rounding is close to zero or even negative, as can be seen from Figs. 9.2 and 9.8. Analytical results presented in Figs. 9.6 and 9.7 can be generalized if instead of

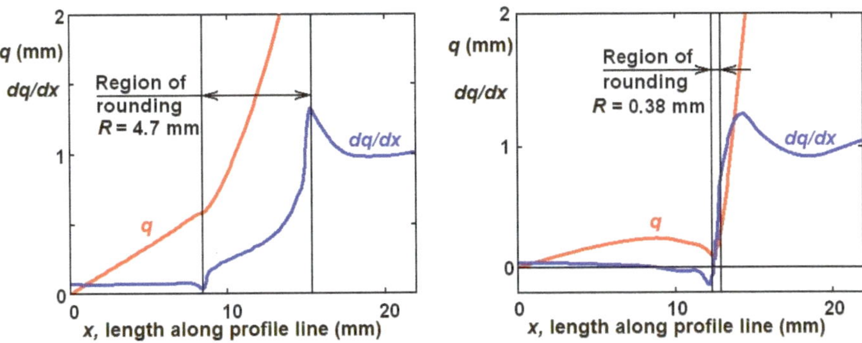

Fig. 9.8 $q(x)$ and dq/dx for corner rounding of 4.7 and 0.38 mm, see geometry in Fig. 9.1. $x = 0$ corresponds to the cavity equator

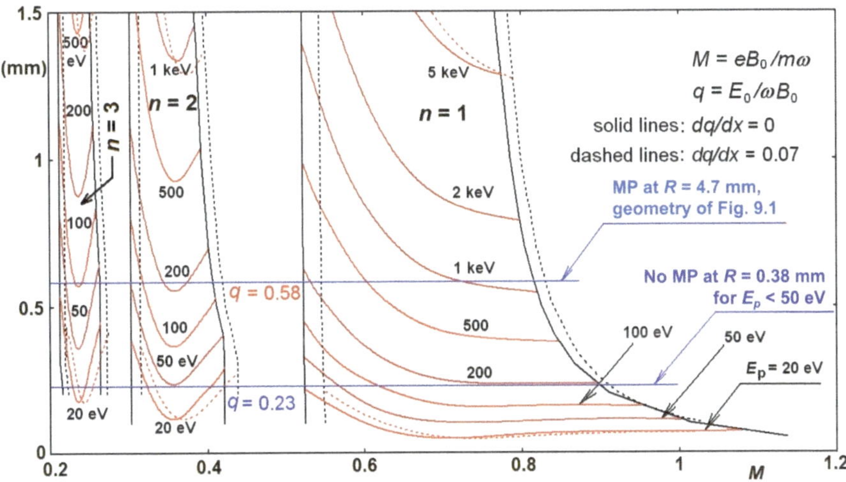

Fig. 9.9 First 3 zones of MP1 with $dq/dx = 0$ and $dq/dx = 0.07$. Emission energy of secondary electrons $E_s = 4\,\text{eV}$; E_p is the impact energy of primary electrons

electric field on axis we use the magnetic parameter M for the surface magnetic field at the equator corresponding to this electric field. Then, a multipactor map can be created, as shown in Fig. 9.9.

In the experiment described in [3], five zones of multipacting were observed (with order n from 2 to 6) for the rounding radius of $R = 4.7\,\text{mm}$. The lowest impact energy was 50 eV. No multipacting occurred when this radius was changed to 0.38 mm. We can see in Fig. 9.9 that the maximal energy of multipacting electrons for the second and third zones ($n = 2$ and 3), and obviously for higher zones, is below 50 eV if the rounding radius is $R = 0.38\,\text{mm}$. The field in the experiment was not high enough to excite the first-order multipactor; the maximal axial field value was below 20 MV/m, which corresponds to $M = 0.42$. Therefore, this generalized description, in terms of geometric (q) and field parameters (M), is also in full agreement with the experiment.

9.5 Phase and Space Stability: Traveling Multipactor

In the calculation of the zone boundaries (see Figs. 9.6 and 9.9), the condition of stability or the condition of return was used, as described above. After each flight, the particles fall on the surface at a new place, shifting from the starting point. This differentiates MP1 from MP2 as both phase and space stability were found for the latter in Chap. 8.

Figure 9.10 (top) illustrates the drift of the impact points for electrons starting at the edge of the flat equatorial region. The calculations were performed for the S-band cavity geometry (Fig. 9.1) with rounding radius $R = 4.7$ mm. In these

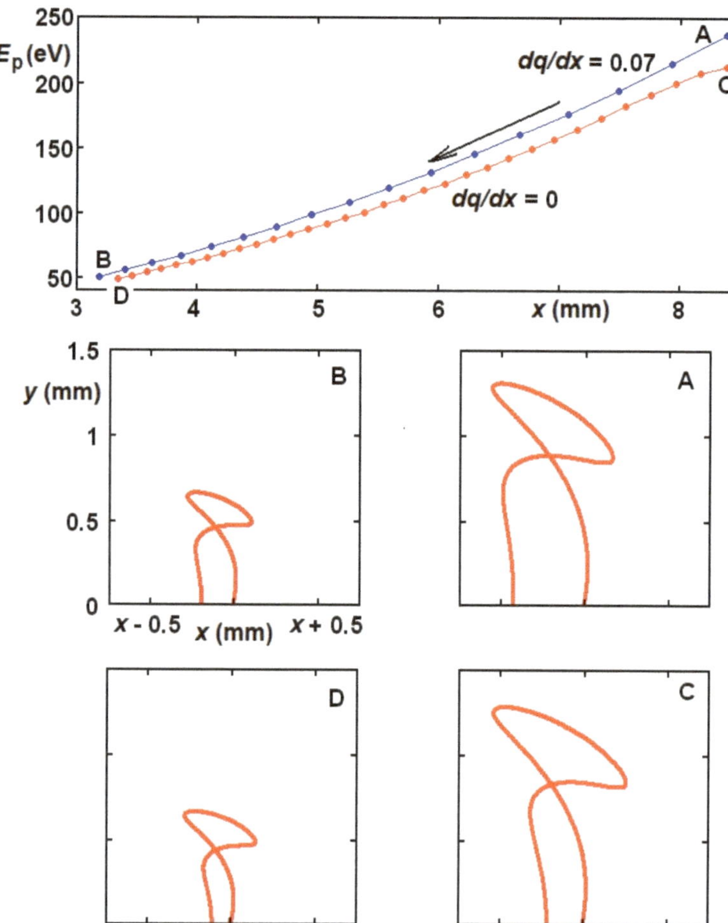

Fig. 9.10 Top: Shift of the impact points for electrons starting at the edge of the flat equatorial region. Two curves correspond to different values of dq/dx. Bottom: Trajectories at the extreme points A, B, C, and D shown on the same scale

calculations, motion of a single particle was traced under assumptions that SEY $= 1$, the secondary particle energy is 4 eV, and its starting velocity is normal to the surface. The lower curve is calculated for $M = 0.35$ (in the middle of the $n = 2$ multipacting zone), and $q = 0.58$ mm (see Fig. 9.9), with $dq/dx = 0$. The calculation results show that the particle can make approximately 30 impacts with $E_p > 50$ eV, so a large number of secondary electrons can be produced. The upper curve is calculated for $M = 0.379$, the same initial q, but with a realistic value of $dq/dx = 0.07$. The values of M are chosen so that the stable phase angle is close to 0. The value of q was recalculated for the lower curve after each impact according to the coordinate shift (Fig. 9.8, left). One can see that accounting for the derivative in (9.8) significantly increases the width of the trajectories (Fig. 9.10), though the boundaries of the zones are only slightly shifted (Fig. 9.9). Nevertheless, the number of impacts – when the electrons are in the region with SEY > 1 – is big enough and multipactor can exist. The phase change after each flight is very small, less than $0.02°$ in all cases, so phase stability is preserved.

The measure of the phase stability, $\lambda = \partial\theta_2/\partial\theta_1$, changes from 0.135 to 0.287 in the case of $dq/dx = 0$ and from 0.134 to 0.280 in the case of $dq/dx = 0.07$. Here, θ_1 and θ_2 are the phases of start and impact respectively. The condition $|\lambda| < 1$ means that a random phase deviation from the equilibrium phase at the start point becomes smaller at the point of impact which is the next start point. The requirement $|\lambda| = 1$ defines the boundaries of the zones, except for the cases of returning particles in the decelerating field. We can conclude that the one-point multipactor is resonant as far as the phase stability is concerned, but the point of impact continuously shifts toward the equator. Therefore, we name this phenomenon *"traveling multipactor"*.

This traveling feature of MP1 was not noticed in [4] as it was masked by many details taken into account, such as distributed start velocities, reflected electrons, and simply because too many trajectories are simultaneously shown in figures. Although these details reflect the real situation, they act like noise, hiding the main features of the phenomenon. Therefore, MP1 *is resonant in nature*, but this is only a phase resonance as the multipactor travels along the cavity profile line toward the equator.

The electric field in the area where the flat region transitions to the rounded corner has low dq/dx, and hence the multipactor travels slowly there. When $dq/dx < 0$ – as in the case of geometry with $R = 1$ mm (see Figs. 9.4 and 9.8) – reverse motion is possible. However, the value of q is too low, meaning the impact energy is too low, so SEY < 1 and multipactor is suppressed. This is why the example presented in Fig. 9.10 with $dq/dx = 0$ is artificial; it should only show that for different q, if there is a "point of rest" – with a small dq/dx, like at the beginning of the rounding – the probability of multipacting increases. For more detail, the influence of the value of dq/dx on the trajectory shift in each flight is shown in Fig. 9.11. Here, the phase angle is also chosen to be zero within the second multipactor zone. The value of q is taken as 0.58 mm. As the "steps" become bigger with the increase of dq/dx, fewer steps are needed to reach the place where $E_p = 50$ eV. Besides, for greater dq/dx, the phase shift between consecutive flights becomes different and the particle goes out of the phase resonance. Thus, geometries with large dq/dx are

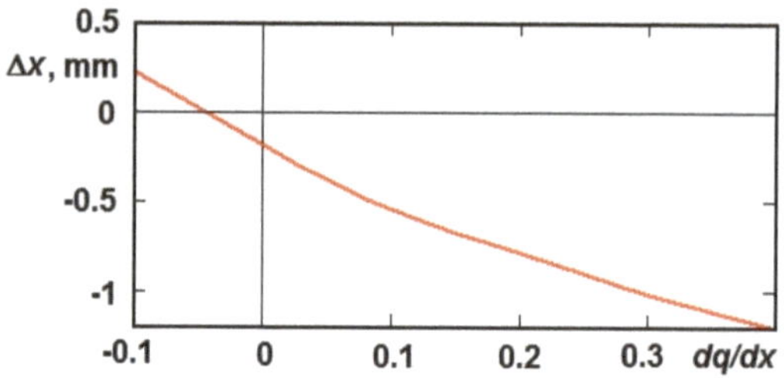

Fig. 9.11 Dependence of shift of the impact point after each flight on the derivative dq/dx

not favorable for supporting MP1. Cavities with a flat equator, like in Fig. 9.1, have low dq/dx and are multipactor-prone.

The geometric parameter p introduced in Chap. 8 is the same as dq/dx. For the S-band cavity under consideration, $p = dq/dx = 0.07$, and hence MP2 does not happen at its equator as p is too low, $\ll 0.3$. On the other hand, the elliptical cavities have big $p = dq/dx$. For example, for the TESLA cavities, this value is 0.29, and for the Cornell ERL cavity it is 0.28 (see Table 8.1), that MP1 is unlikely in such cavities.

Examples of the orbits for $q = 0.58$, $dq/dx = 0.07$ at the beginning, in the center, and at the end of the first three zones are shown in Fig. 9.12. One can see that the sizes of orbits do not differ too much in spite of different values of fields and different start phases (θ_1); all pictures have the same scale. The energy of impact (E_p) of the primary electrons is also indicated in each plot. The start of the secondary electrons in a stable motion can occur at different phase angles, not only at $\theta_1 = 0$. However, for higher order zones, starting from the second, maximal energy and accordingly maximal SEY happen at the angle close to zero, which also follows from Fig. 9.9 (minima of curves E_p = const are in the middle of zones $n = 2$ and 3).

It is worth noting that the sizes of orbits do not scale with the size of the cavity but remain the same because the initial energy of the secondary electrons remains the same at approximately 2 to 4 eV. This means that the presented approach will work better for bigger cavities of the same shape because the assumption that the sizes of orbits are smaller than the curvature radius will be better satisfied.

9.6 Comparison of the Equations of Motion for MP1 and MP2

Equations of motion for the two-point multipactor were derived as (8.15) in Sect. 8.5. The geometrical parameter of the two-point multipactor is defined as $p = (dE_s/dx)/\omega B_s$. The difference in signs between (9.8) and (8.15) on the right sides means only a shift of the origin of the phase angle θ by the value of π. If we take $q = px$, the sets (9.8) and (8.15) coincide, correct to the phase shift by π. The difference in solutions results from the fact that in the case of MP2 the particle

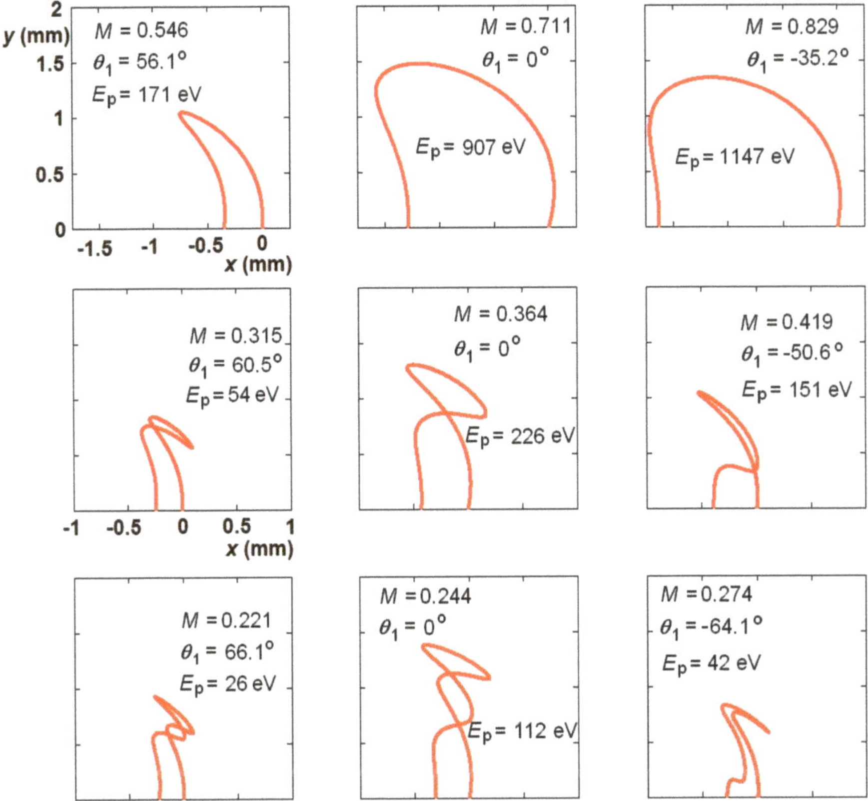

Fig. 9.12 Examples of orbits for the first 3 zones of MP1. The top row is for $n = 1$, the middle row is for $n = 2$, and the bottom row is for $n = 3$. Compare values of M at $q = 0.58$ with Fig. 9.9

changes the sign of x each time it flies over the equator, but in the case of MP1, the particle is at the same side of the equator moving closer to it step by step. On the map of MP2 (e.g., see Fig. 8.8), only one point corresponds to a given value of the field parameter M because the value of p is defined at one point only, the equator. In the case of MP1, each value of M corresponds to a line on the map of Fig. 9.9 because different q values can be found on the profile line. Particles make big steps along this line at higher q values, can slow down at low derivatives dq/dx, and, if the number of steps with SEY > 1 is big enough, multipactor occurs.

The equations of (8.15) are homogeneous, that is, if all the functions – x, y, and their derivatives – are multiplied by a constant, the equations hold true. For example, the sizes of the trajectories as well as velocities and accelerations at each point of the trajectory increase by $\sqrt{2}$ times if the emission energy is doubled. Conditions of stability will also hold if the emission velocity changes.

This is also true in the case of MP1 if the value of $q(x)$ in (9.8) is proportional to x. Moreover, while in the case of MP2, the changing of initial velocity around the mean value will cause a deviation from the mean trajectory, which will decrease at the next flight due to phase stability, in the case of MP1, the phase will be stable if the velocity changes, and only the size of the step to the equator will change.

The value of p in (8.15), as shown in Chap. 8, does not depend on frequency but only on the shape of the cavity. The same is true for MP1 if q is proportional to x. This means that the map of MP1 as well as the map for MP2 can be used for any frequency.

From the linearity of the equations of motion and previous considerations, it follows that the impact energy of primary electrons E_p is proportional to the energy of secondary electrons. For example, if the secondary electrons have a mean value of energy of 2 eV, not 4 eV, all values of E_p in Fig. 9.9 should be divided by 2.

The angle at which the electrons impinge on the surface can change significantly in the case of MP2, increasing the surface emission yield. This angle is zero for MP1, as follows from (9.5) for $q = \text{const}$, but it also appears small ($<5°$) for the discussed geometry with $dq/dx = 0.07$.

9.7 Conclusion

In this chapter, we presented an explicit description of the one-point multipactor. It was done following a similar description of two-point multipacting in Chap. 8.

Small sizes of trajectories in MP1 require a very precise calculation of fields for simulations. In the proposed approach, due to these small sizes, fields are presented as Taylor expansions, and trajectories are found solving ordinary differential equations of motion. Conditions of the phase stability define the boundaries of the multipactor zones. The MP1 discharge – similar to MP2 – is described in terms of two parameters: geometrical parameter and field parameter. The geometrical parameter p at the equator or q along the profile line of the cavity and the value of the field parameter M at these points define the conditions when one- or two-point multipactor will occur. A comparison with some published results showed that this approach gives us very good agreement with simulations and experimental data. The multipactor maps provide a convenient way to quickly evaluate boundaries of potential multipacting zones and direct follow-up studies with computer codes that could be time-consuming without prior insight.

The past studies on multipacting in SRF cavities have led to the adoption of elliptical shape for high-β resonators [7, 8], which do not support MP1 as the electrons quickly travel toward the equator, where MP1 is not sustained. However, the general approach presented in Chaps. 8 and 9 – involving the concept of geometrical and field parameters – is also applicable to other RF resonators, including normal conducting accelerating cavities, low-β resonators (quarter-wave, half-wave, and spoke resonators), deflecting cavities, and a wide variety of other shapes.

References

1. V. Shemelin, Conditions for the existence of 1- and 2-point multipactor in SRF cavities, in *Proceedings of IPAC2013, International Particle Accelerator Conference* (Shanghai, 2013), p. 2456
2. V. Shemelin, One-point multipactor in crossed fields of rf cavities. Phys. Rev. Acc. Beams **16**, 102003 (2013)
3. C.M. Lyneis, H.A. Schwettman, J.P. Turneaure, Elimination of electron multipacting in superconducting structures for electron accelerators. Appl. Phys. Lett. **1**(8), 541 (1977)
4. I. Ben-Zvi, J.F. Crawford, J.P. Turneaure, Electron multiplication in cavities, in *Proceedings of PAC1973, Particle Accelerator Conference* (San Francisco, 1973), p. 54
5. H. Padamsee, A. Joshi, Secondary electron emission measurements on materials used for superconducting microwave cavities. J. Appl. Phys. **50**(2), 1112 (1979)
6. https://www.euclidtechlabs.com/superlans. Accessed 8 Jul 2020
7. U. Klein, D. Proch, Multipacting in superconducting RF structures, in *Proceedings of Conference on Future Possibilities for Electron Accelerators* (Charlottesville, 1979), p. N1
8. P. Kneisel, R. Vincon, J. Halbritter, First results on elliptically shaped cavities. Nucl. Intsrum. Methods Phys. Res. A **188**, 669 (1981)

Part III
Multipacting-Free Cavities and Transitions Between Cavities and Beam Pipes

Chapter 10
Optimized Shape Cavities Free of Multipacting

10.1 Introduction

Most of thesuperconducting radio-frequency cavities in operation today belong for speed-of-light particles to the same class elliptical cavities. In this chapter, we consider how the process of elliptical cavity optimization can be utilized in creating cavity shapes free of multipacting.

A superconducting cavity should be properly optimized. Typically, optimization involves minimizing peak surface electric and magnetic fields [2]. For cavities designed to operate at high accelerating gradients E_{acc}, the limiting factor is the maximal (or peak) surface magnetic field H_{pk}. Therefore, minimization of the value of H_{pk}/E_{acc} is the first task for cavity optimization, which is closely related to the decrease of RF losses in the cavity walls.

In this chapter, we describe the cavity optimization process and demonstrate that some substantial deviations of the shape can be made without deterioration of H_{pk}/E_{acc}. This opens a way to overcome multipactor (as well as some other problems) in elliptical cavities and thus design multipactor-free cavities.

10.2 Method of Optimization

The optimization described below is enabled by accurate calculations of cavity parameters. 2D computer code SuperLANS by D. Myakishev and V. Yakovlev [3] and the TunedCell envelope code by D. Myakishev [4] have the accuracy necessary for our optimization [5].

The chapter is based on materials first published in [1].

V. D. Shemelin, S. A. Belomestnykh, *Multipactor in Accelerating Cavities*, Particle Acceleration and Detection, https://doi.org/10.1007/978-3-030-48198-8_10

From the optimization reported in [6], we know that the minimum H_{pk}/E_{acc} is a monotone function of E_{pk}/E_{acc} – it is decreasing with increasing E_{pk}/E_{acc} – and a monotone function of the wall slope angle α – it is decreasing as the angle becomes smaller.

From this point on, we will use $h = H_{pk}/E_{acc}/42$ and $e = E_{pk}/E_{acc}/2$ instead of H_{pk}/E_{acc} and E_{pk}/E_{acc}, respectively, for the reasons described in Sect. 6.2.

Therefore, the goal of the cavity optimization is to find the minimum of h under given constraints: the value of e should be less than or equal to the maximal allowed, and the value of α should be greater than or equal to the minimal allowed. We are going to minimize the function $h(A, B, a, b)$ under these conditions. The “brute force”, or “grid search” approach is to calculate h on a 4-dimensional grid around an initial central point, find the minimum of h between the neighbouring points, make the point of minimum h the new central point, and repeat the procedure until the minimum of h is found for the given grid step size. Then, the value of the step can be decreased until we reach the required accuracy. However, a 2D example presented in Fig. 10.1 shows that this method is inefficient when the optimized function belongs to a class of so called ravine functions. More advanced methods, such as gradient descent, also appeared inefficient in our case. “Gradient descent becomes difficult when the function being optimized looks locally like a ravine” [7]. At the points marked with crosses around the point A (Fig. 10.1), the level of the surface is higher than at the point A. This level is decreasing to points B, C, and D. In this example, steps along Y should be two times bigger than in the direction X to find the correct minimum.

To prove that the cavity optimization problem belongs to the ravine type, we tried to solve the optimization problem with the Levenberg-Marquardt algorithm [8],

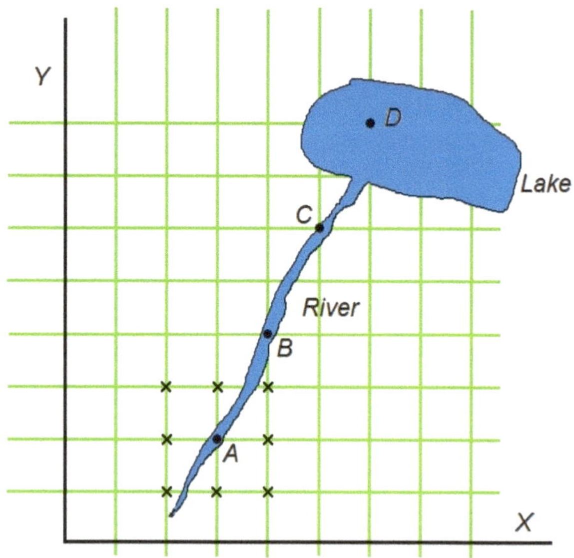

Fig. 10.1 Brute force optimization does not lead to the minimum

which is gradient-based. In fact, as predicted by the literature, due to the ravine shape of the problem, it did not show good convergence compared to a Monte-Carlo-based algorithm. We used the Monte Carlo scheme that relies on pre-computed values, which is often referred to as the Markov chain Monte Carlo algorithm in the literature [9]. The Monte Carlo method is faster than the grid search method, but it also gets "stuck" very often, not reaching the ultimate minimum.

As will become clear below (Figs. 10.3 and 10.4), we have exactly the case of a ravine function: a rapid descent to the boundary and gently declined path along the boundary. In the search of the minimum h, the algorithm quickly comes to the boundary where the limiting values of e and α are reached. To move along the "bottom" of the ravine, we can solve the following system of equations:

$$\begin{aligned} \partial h/\partial A \cdot \Delta A + \partial h/\partial B \cdot \Delta B + \partial h/\partial a \cdot \Delta a + \partial h/\partial b \cdot \Delta b &= \Delta h, \\ \partial e/\partial A \cdot \Delta A + \partial e/\partial B \cdot \Delta B + \partial e/\partial a \cdot \Delta a + \partial e/\partial b \cdot \Delta b &= \Delta e = 0, \\ \partial \alpha/\partial A \cdot \Delta A + \partial \alpha/\partial B \cdot \Delta B + \partial \alpha/\partial a \cdot \Delta a + \partial \alpha/\partial b \cdot \Delta b &= \Delta \alpha = 0, \end{aligned} \quad (10.1)$$

where, Δh is a negative value to be added to h to decrease it.

The system is underdetermined as we have three equations and four unknowns: ΔA, ΔB, Δa, and Δb. A fourth equation can be added to define length of the vector of increment, s, so that the system of equations becomes solvable:

$$\Delta A^2 + \Delta B^2 + \Delta a^2 + \Delta b^2 = s^2. \quad (10.2)$$

For any Δh, we can choose s, preferably small, and solve this system of equations. Therefore, we can consider this approach as an implementation of the gradient method in the case of the ravine function and the function under some constraints.

We can solve the system of equations (10.1) and (10.2) for any negative value of Δh and find a lower point than the point we started at. It looks like we can decrease h for indefinitely long time. Of course, this is possible only as long as the absolute value of Δh is small enough that the partial derivatives do not change significantly.

Nevertheless, it is not clear from this system that the minimum is reached. It becomes clear if we consider the dependencies of h or e on A, B, a, and b. In Fig. 10.2 the dependencies of h and e on a and b are shown for the case of $\beta = 1$, $R_a = 35$ mm, $e = 1$, and $\alpha = 100°$. The minimum h appears near the point of break of the dependencies. When the derivative is considered for the left side of the point of break ($\Delta a < 0$), the resulting solution of the system (10.1) and (10.2) gives $\Delta a > 0$, and vice versa, considering $\Delta a > 0$, the solution gives $\Delta a < 0$. This contradiction shows that further decrease of h is impossible, so the minimum is reached.

The same consideration can be made for all other partial derivatives in (10.1). However, in practice, we observed only derivatives of h and e with respect to a and b changing their behavior near the point of minimum h.

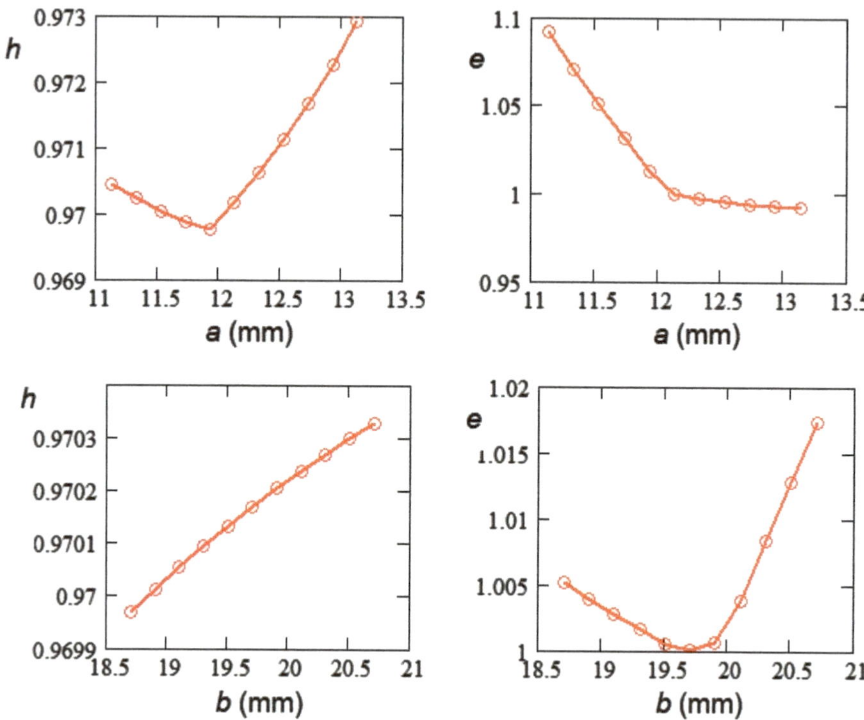

Fig. 10.2 Dependencies of h and e on a and b near the point of min h

The cause of this break is as follows: for an optimal cavity shape, there are two local maxima on the curves of H vs L/L_0. One can see them in Fig. 6.7 for LL and RE cavities. The first maximum is at $L = 0$, and the second is in the right part of each curve, e.g., near the point A for the RE cavity. The TESLA cavity is slightly under-optimized and has only one maximum, at $L = 0$. There can also be two local maxima on the flat part of the curve E vs L/L_0 (Fig. 6.6). The second local maximum E at $L/L_0 = 1$ appears by thorough optimization and practically cannot be seen on the flat part of curve 2 in Fig. 6.6. These maxima are not very well pronounced on these graphs because the curves become very flat between them in the process of optimization. When changing, for example, the half-axis a, trying to minimize H_{pk} (and therefore h) we move to the minimum of h, the bigger maximum of H decreases but the smaller one grows and can overtake the other. At the point of break (Fig. 10.2), these maxima become equal, and if we continue to change a, the value of H_{pk} is defined now by the other hump, and its dependence on a changes. Therefore, the minimum h is reached when the humps on the curve H vs L/L_0 or on the curve E vs L/L_0 become equal in height. In the last case, the decrease of $h = H_{pk}/E_{acc}$ due to equalization of local maxima E becomes not because of minimization of H_{pk} but because of an increase of E_{acc} when the force lines are reaching their maximum uniformly around the iris tip.

The optimization procedure described above can lead to shapes that are not realizable for inner cells of a multi-cell cavity due to space constraints. In addition, SRF cavities are usually electron-beam-welded from half-cells stamped from niobium sheets [10], which imposes additional restrictions. Therefore, for the reentrant cavity, the value of A cannot be greater than the cell length. Moreover, considering the thickness of the wall t, usually 3 mm, the restriction $A+t < L$ should be applied. If we consider the necessary gap between cells needed for welding them together, we should make this restriction even more severe. The gap $L - (A + t)$ should be several millimeters wide to make welding of the cells possible.

To guarantee accuracy in the process of stamping the half-cells, the curvature radius of the iris cannot be too small even if it doesn't increase the E_{pk}. A reasonable minimal value of this radius is twice the thickness of the niobium sheet used in fabrication, i.e., $r_c > 2t$. For example, the TRASCO-ASH cavity for $\beta = 0.47$ and frequency $f = 704.4$ MHz [11] has the radius of curvature at the iris tip of $r_c = 6.1$ mm.

10.3 Optimization of the TESLA Cavity

The TESLA cavity [12, 13] has been thoroughly studied because it was designed for the International Linear Collider and is currently used in many accelerators throughout the world. In spite of "round" values of the elliptical half-axes ($A = B = 42$, $a = 12$, $b = 19$ mm), its shape is very close to the optimized one. Fig. 10.3 presents contour plots of the normalized parameters h and e and the wall slope angle α for fixed values of $a = 12$ mm and $b = 19$ mm on the $A - B$ plane around the nominal point. The pink area for e and blue area for α are "forbidden" areas, i.e., areas beyond the limiting values of $e = 0.997$ and $\alpha = 103.17°$. We call these areas "forbidden " in the sense described in the previous section: the goal of optimization is to find the minimum h for e less than or equal to a given maximal value and α greater than or equal to a given minimal value, so the values e and α violating these inequalities are "forbidden". The lower right-hand side picture shows contours of h overlaid with forbidden areas, the "twice forbidden" area is colored green. One can see that there is a possibility to improve the value of h from $h = 0.99$ by moving down along the pink area boundary.

After further optimization, we come to Fig. 10.4. Here, we slightly deviated from the TESLA geometry by choosing $\alpha = 100°$, which is less than in the original cavity (103.17°), and $e = 1$. Both of these changes improve the value of h from 0.99 to 0.97, but this is not the main result.

We believe that the main result is the fact that by moving along the forbidden area boundary, we can change both A and B in a wide range while changing the value of h by no more than 0.5%. As will be shown later, this very shallow minimum gives us the possibility to create a practically optimal cavity geometry free of multipactor.

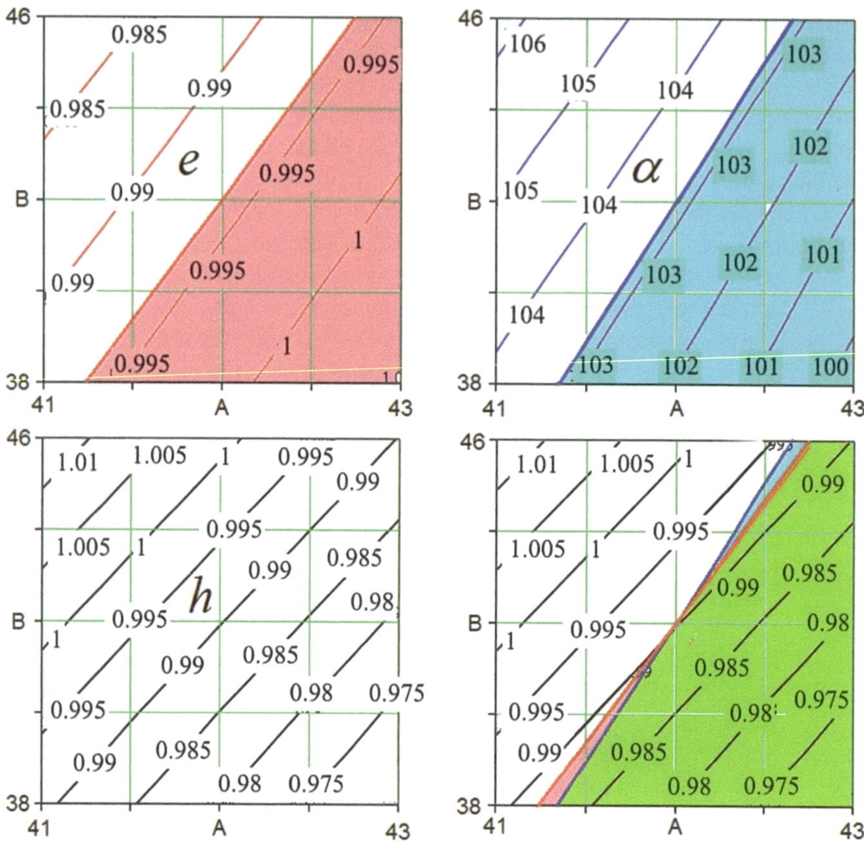

Fig. 10.3 Contour plots for TESLA cavity near the nominal values of A and B. $a = 12$ mm and $b = 19$ mm

10.4 Multipactor Considerations

In elliptical superconducting cavities, multipactor occurs most often at the cavity equator, where it is a two-point multipactor. The method of predicting the appearance of this type of multipactor without time-consuming simulations but immediately from the field calculations done with a 2D program was presented in Chap. 8.

Multipacting zones at the cavity equator (Fig. 8.8) are defined by two coefficients: geometrical coefficient $p = \beta/(\alpha + \beta)$ and field coefficient $M = eB_0/m\omega$. α and β are factors of proportionality for the longitudinal and radial components of the electric field near the equator, $E_z = \alpha d$ and $E_r = \beta d$, respectively. The points of definition of these factors are at a distance $d = 1$ mm (Fig. 10.5), a characteristic size for the orbit of an electron in this kind of multipactor at any frequency, as

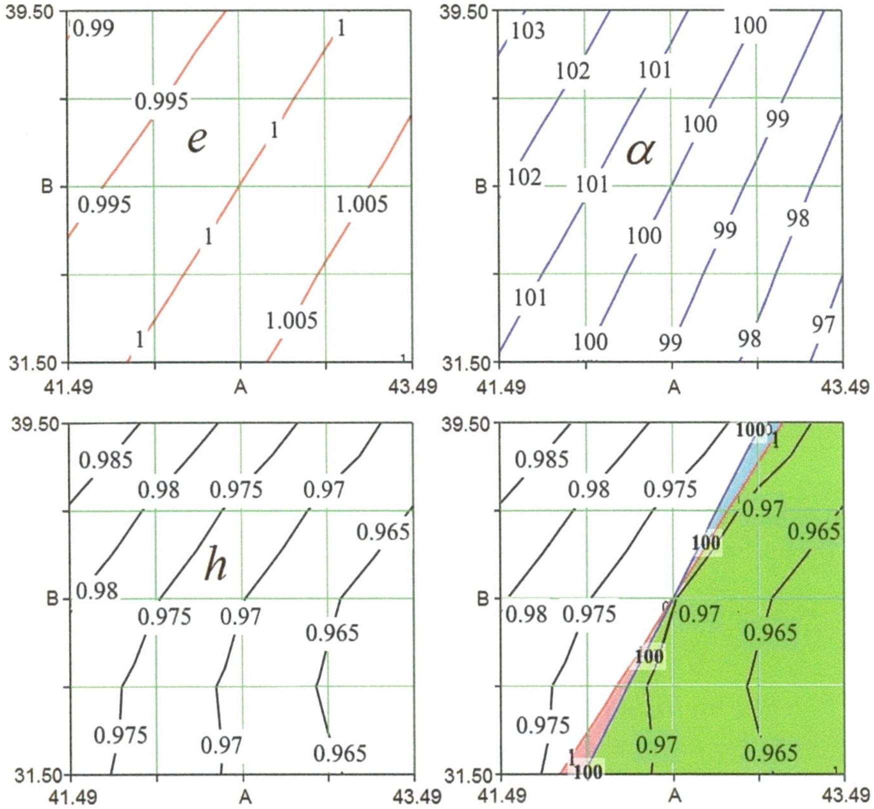

Fig. 10.4 Contour plots for the optimized TESLA cavity with $a = 12.13$ mm, $b = 19.70$ mm

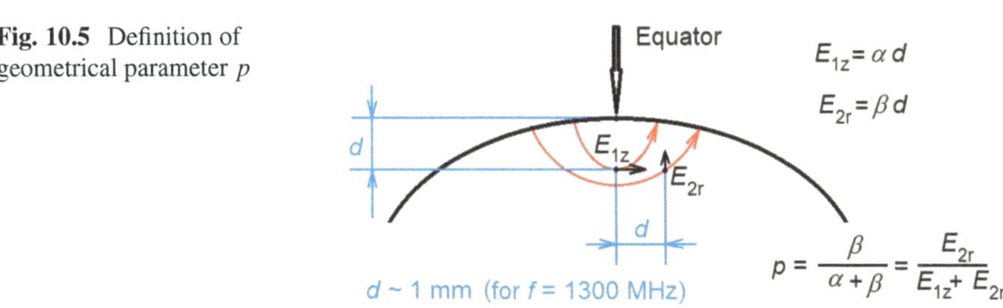

Fig. 10.5 Definition of geometrical parameter p

discussed in Chap. 8. Fields E_r and E_z at these two points can be found by any 2D electromagnetic solver.

B_0 in the expression for the field coefficient M is the amplitude of the magnetic field at the equator. B_0 is proportional to the accelerating field E_{acc} in the cavity and is usually equal or close to the maximal magnetic field in the cavity, B_{pk}.

Each cavity shape has its own geometrical coefficient p. Moving along the line $p = \text{const}$ on the multipactor map shown in Fig. 8.8, we increase the field coefficient

M and hence the magnetic filed at the equator. When the energy of the primary electrons E_p (see the graph) becomes high enough for the SEY to exceed unity, the multipactor starts. This usually happens near $M = 2$, as can be seen from the map. The results for different cavities presented in Fig. 8.8 show that the critical value for multipactor is approximately $p = 0.3$, so we have a very weak and usually easily processed multipactor in the TESLA cavity ($p = 0.286$), a well pronounced multipactor in the Mark I cavity fabricated specially for studying this phenomenon ($p = 0.303$), and practically no multipactor at the equator of the Cornell ERL cavity ($p = 0.276$).

The force lines of the electric field near the equator have elliptical arc shapes (Fig. 10.6) as follows from

$$\frac{E_r}{E_z} = \frac{dr}{dz} = \frac{-\beta z}{\alpha r}, \tag{10.3}$$

where r is measured from the equator toward the axis, i.e., opposite to R, z coincides with Z, as shown in Fig. 10.6.

The solution of this differential equation is

$$\frac{z^2}{\alpha} + \frac{r^2}{\beta} = \text{const.} \tag{10.4}$$

Therefore, the ratio of the half-axes of the ellipic force lines near the equator is $b_{fl}/a_{fl} = \sqrt{\beta/\alpha} = \sqrt{p/(1-p)}$, and the condition $p > 0.3$ corresponds to $b_{fl}/a_{fl} > \sqrt{3/7} \approx 0.655$.

As stated above in the Sect. 10.3, we can change the dimensions A and B so that the value of normalized magnetic field h will only slightly change in a wide

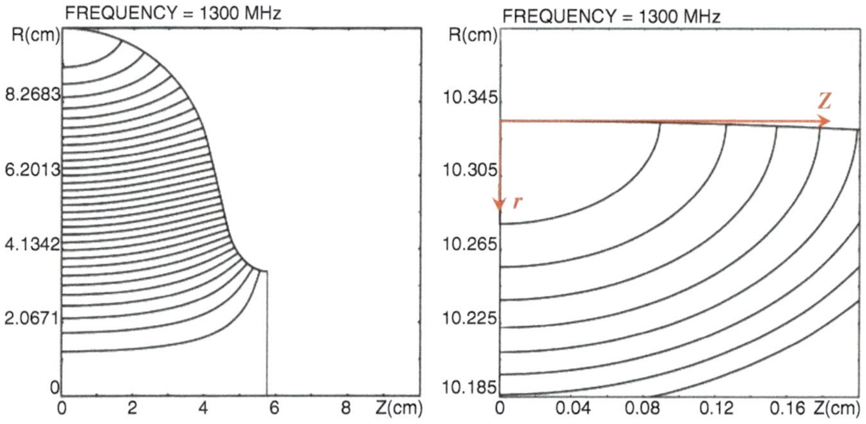

Fig. 10.6 Electric field force lines for the entire half-cell of the TESLA cavity (left) and near the equator (right). Force lines near the equator are elliptic arcs with ratio of half-axes $b_{fl}/a_{fl} = 0.633 < 0.655$; see text

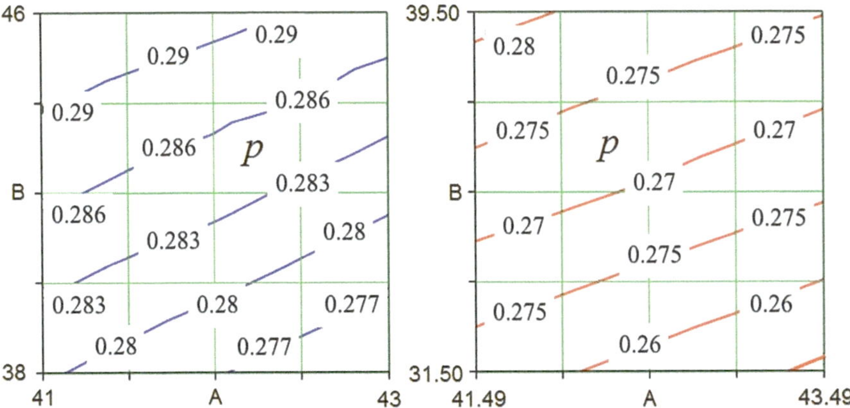

Fig. 10.7 Parameter p for the TESLA cavity: original (left) and optimized (right) in the same coordinates as in Figs. 10.3 and 10.4

range of A and B. For the example presented in Figs. 10.3 and 10.4, let us present values of p in the same coordinates, as shown in Fig. 10.7. One can observe that (i) the optimized cavity has smaller p, so there will be no multipacting, and (ii) this parameter can be further decreased by allowing h to be increased.

10.4.1 Multipactor in the End Cells of a Multicell RF Cavity [14]

As it was shown in [15], see also Chap. 11, multipactor electrons are attracted to the electric field minimum on the cavity surface. This minimum may shift off the middle plane if the cell is asymmetric, for example, if it has an inner cell on one side and a beam pipe on the another side (an end cell of multi-cell cavity). This shifted minimum can be treated in the same way as the minimum on the equator. The changing curvature of the surface near this point can be neglected because the distance at which the fields have to be calculated are usually much smaller than the curvature radius at this point. The only complication is the necessity to draw a perpendicular line to that point on the ellipse, while it is straightforward to draw this line from the equator.

Consider the minimum of electric field shifted to point 0 in Fig. 10.8. We need to erect a perpendicular line to this point on the ellipse and find the fields at a distance d, at point 1, and at points 2 and 3 on a straight line parallel to the tangent at point 0, at a distance d from point 1. Only the upper part of the cell is shown in Fig. 10.8. It is defined by an ellipse with half-axes A and B. We will designate coordinates of points 0, 1, ... as (Z_0, R_0), (Z_1, R_1), ...

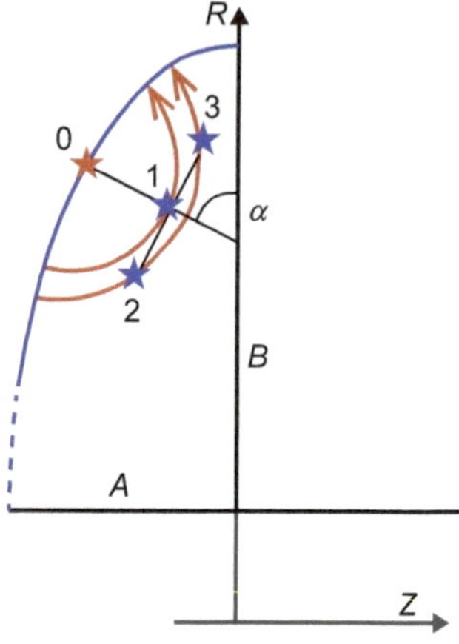

Fig. 10.8 Geometry for the case when the minimum of the electric field is shifted from the equator

Now we can write a system of equations to find coordinates of point 1:

$$\frac{R_1 - R_0}{Z_1 - Z_0} = \frac{A \cdot (A^2 - Z_0^2)^{0.5}}{BZ_0},$$

$$d^2 = (R_1 - R_0)^2 + (Z_1 - Z_0)^2,$$

and of point 2:

$$\frac{R_2 - R_1}{Z_2 - Z_1} = \frac{BZ_0}{A \cdot (A^2 - Z_0^2)^{0.5}},$$

$$d^2 = (R_2 - R_0)^2 + (Z_2 - Z_0)^2.$$

Coordinates of point 3 can be found if index 2 in the latter system changes to 3. Electromagnetic solvers usually provide E_r and E_z components of the electric field. We need to have field components parallel or normal to the tangent at point 0. Let us call them analogously to the fields in the definition of p as E'_{2r} and E'_{1z}. This transformation looks as follows:

$$E'_r = E_r \cos\alpha - E_z \sin\alpha,$$

$$E'_z = E_r \sin\alpha + E_z \cos\alpha.$$

10.4.2 Example: End Cells of the TESLA Cavity

Let us, as an example, calculate the geometrical parameter p for the end cell of the TESLA cavity. The TESLA cavity is asymmetric: it has different end cells that

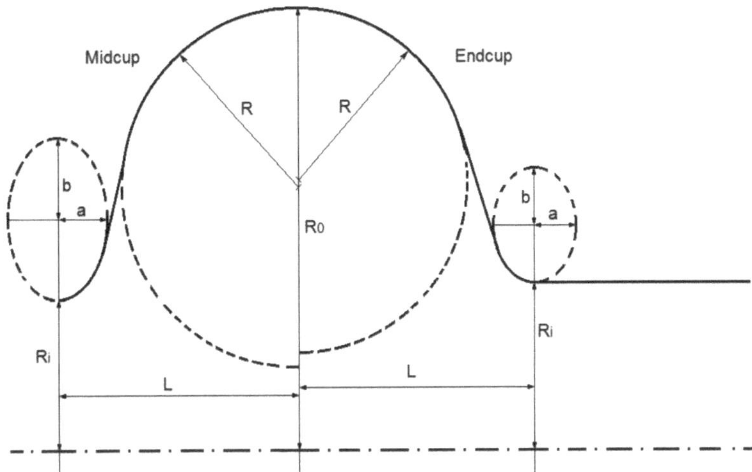

Fig. 10.9 Geometry of the TESLA end cells

Table 10.1 Dimensions (in mm) of the TESLA cavity

Parameter	Midcup	Endcup1	Endcup2
Equatorial radius R_0	103.3	103.3	103.3
External curvature radius R	42	40.34	42
Iris radius R_i	35	39	39
Horizontal half-axis a	12	10	9
Vertical half-axis b	19	13.5	12.8
Length L	57.692	56	57

makes possible to extract some higher order modes otherwise trapped in the inner cells. Consider the end cell 1 with dimensions presented in [16], see Fig. 10.9 and Table 10.1. Actually, these dimensions correspond to the frequency f = 1301.369 MHz instead of 1300 MHz, as was found with SLANS [3]: the older program URMEL did not guarantee a necessary accuracy. However, this fact does not change the results substantially. If we calculate p at point 1 (Figs. 10.5 and 10.8) located on the normal to the equator, we obtain $p_2 = 0.252$ and $p_3 = 0.317$ for points 2 and 3, respectively. This is an indication that the surface electric field minimum is not at the equator. Zero of the surface electric field E_s is shifted to the left from the equator by $\Delta Z = -0.154$ mm, so that the angle $\alpha = \arctan(Z_1 - Z_0)/(R_0 - R_1) = 0.218$. However, even taking into account this small shift gives us almost equal p's, 0.285 and 0.287 for points 2 and 3, respectively. The field calculations were performed with SLANS on a 112 × 86 mesh.

Calculations for the end cell 2 give a shift $\Delta Z = -0.195$ mm and $\alpha = 0.266$ for the minimum of E_s. On the normal to the equator, we can find $p_2 = 0.240$ and $p_3 = 0.327$, but after the correction to the offset point we get $p_2 = 0.285$ and $p_3 = 0.286$. The calculation was repeated with a twice denser mesh (224 × 172), resulting in the shift of $\Delta Z = -0.205$ mm, and $p_2 = p_3 = 0.286$ with an accuracy of three decimal places. For the inner cells consisting of two midcaps, p = 0.286 as was found on the straight line in the plane of the equator, Fig. 10.5. It is known that multipacting in the TESLA cavities is weak and can be easily processed [17]. Note that the fields in formulas for E_r' and E_z' should be taken with the signs as calculated, but in formulas for p, shown in Fig. 10.5, absolute values of vector components should be taken.

10.5 Conclusion

In this chapter, we demonstrated that it is possible to maintain the minimal H_{pk}/E_{acc} while re-optimizing the shape of the cavity. This re-optimization allows one to reduce the geometrical parameter p and thus bring the cavity outside the multipactor zone boundary, creating cavities either less susceptible to multipacting or completely free of multipactor.

References

1. V. Shemelin et al., Systematical study on superconducting radio frequency elliptic cavity shapes applicable to future high energy accelerators and energy recovery linacs. Phys. Rev. Accel. Beams **19**, 102002 (2016)
2. V.D. Shemelin, G.H. Hoffstaetter, First-principle approach for optimization of cavity shape for high gradient and low loss, in *Proceedings of IPAC2012, the Third International Particle Accelerator Conference*, New Orleans, 2012, p. 3015
3. D.G. Myakishev, V.P. Yakovlev, The new possibilities of Superlans code for evaluation of axisymmetric cavities, in *Proceedings of PAC1995, Particle Accelerator Conference and International Conference on High-Energy Accelerators*, Dallas, 1995, p. 2348. https://www.euclidtechlabs.com/superlans. Accessed 8 Jul 2020
4. D. Myakishev, TunedCell, LEPP Report SRF/D051007-01, Cornell University (2005)
5. S. Belomestnykh, Spherical cavity: analytical formulas. Comparison of computer codes. LEPP Report SRF941208-13, Cornell University (1994). https://www.classe.cornell.edu/public/SRF/1994/SRF941208-13/
6. V. Shemelin, High-gradient and low-loss SC cavities with different wall slope angles. LEPP Report SRF070614-02, Cornell University (2007). https://www.classe.cornell.edu/public/SRF/2007/SRF070614-02/Cells_for_HG-LL-home.pdf
7. D. Duvenaud, D. Maclaurin, Citation from Harvard University course CS281: Advanced Machine Learning, Section 3: Practical Optimization
8. J.J. Moré, *The Levenberg-Marquardt Algorithm: Implementation and Theory. Numerical Analysis* (Springer, New York, 1978) p. 105–116
9. W.R. Gilks, *Markov Chain Monte Carlo* (Wiley, New York, 2005)
10. H. Padamsee, J. Knobloch, T. Hays, *RF Superconductivity for Accelerators* (Wiley, New York, 1998). ISBN 0-471-15432-6
11. C. Pagani et al., Design criteria for elliptical cavities, in *Proceedings of 10th Workshop on RF Superconductivity*, Tsukuba, 2001, p. 115
12. TESLA Test Facility Linac – Design Report. Editor D. A. Edwards. DESY Print, TESLA 95-01 (Mar 1995)

13. B. Aune et al., Superconducting TESLA cavities. Phys. Rev. ST Accel. Beams **3**, 092001 (2000)
14. V. Shemelin, Multipactor in the end cells of a multicell rf cavity. Phys. Rev. Accel. Beams **26**, 082001 (2023)
15. S. Belomestnykh, V. Shemelin, Multipacting-free transitions between cavities and beam-pipes. Nucl. Instrum. Methods Phys. Res. A **595**, 293 (2008)
16. D. Proch, The TESLA Cavity: Design Considerations and RF Properties, in *Proceedings of the Sixth Workshop on RF Superconductivity*, CEBAF, Newport News, Virginia, USA, 1993, p. 382
17. K. Twarowski, L. Lilje, D. Reschke, Multipacting in 9-cell TESLA cavities, in *Proceedings of the 11th Workshop on RF Superconductivity*, Lübeck/Travemünde, Germany, 2003, p. 733

Chapter 11
Multipacting-Free Transitions Between Cavities and Beam Pipes. Theorem of Minimal Electric Field

11.1 Introduction

Multipacting in an elliptic superconducting cavity can occur near the equator, but it is usually not very strong, and as shown in Chaps. 8 and 10, one can eliminate this multipacting by changing the cell shape in the equatorial region. It was thought that such cavities are essentially multipactor-free. However, tests of the Cornell Energy Recovery Linac (ERL) injector cavity [2] and KEK Ichiro cavity [3] showed relatively strong multipacting, which was later attributed through computer simulations to the transition regions between the cavity end cells and beam pipes. For the ERL cavity, the presence of multipacting in this region was experimentally confirmed by measuring electric current at a biased probe and correlated temperature changes at the outer wall [2]. Our analysis shows that the amplitude of the electric field along the cavity profile line has a minimum at the location of the multipactor discharge. Similarly, the possibility of multipactor existence in the transition region was found during our work on designing a multicell superconducting cavity for the future Cornell ERL main linac.

We proposed an explanation that an electric field minimum is associated with the local RF potential well, thus attracting electrons to its location and creating conditions favorable to multipacting. Simulations with the computer code MultiPac [4, 5] confirm that smoothing out the transition to eliminate the minimum results in suppression of multipacting.

Material of this chapter was first published in [1].

V. D. Shemelin, S. A. Belomestnykh, *Multipactor in Accelerating Cavities*, Particle Acceleration and Detection, https://doi.org/10.1007/978-3-030-48198-8_11

11.2 Cavity with Transition from Iris to a Larger Diameter Beam Pipe

The transition from the cavity end cell to a beam pipe is shown in Fig. 11.1. The contour line of this transition consists of elliptic arcs connected with tangential straight segments. A_e, B_e, A_i, B_i, and so on are half-axes of the ellipses, where i refers to the inner half of the cell, e refers to the outer half, R is the radius of the circle smoothening the transition, R_{eq} is the equatorial radius, and R_{bp} is the radius of the beam pipe. We examine transitions from the end iris aperture $R_{ae} = 37$ mm to the beam pipe radius $R_{bp} = 55$ mm with different radii R. Half-axes of the end iris ellipses are $a_e = a_t = 12.53$ and $b_t = b_e = 20.95$ mm. Other dimensions of the cavity are chosen to tune its frequency to 1300 MHz and the ratio of the peak electric field to accelerating field to $E_{acc}/E_{pk} = 2.0$.

Results of multipacting simulations for $R = 9$ mm are presented in Fig. 11.2. We can see nearly stable trajectories with the final impact energy mainly sufficient for multiplication. However, there are some irregularities in the phase motion, which repeatedly cause a low impact energy and the cumulative SEY becomes less than 1. The enhanced counter function [5] e_{20}/c_0, as designated on the graph, is approximately 0.1, i.e., less than unity, which is necessary for a sustained discharge. In fact, e_{20}/c_0 is a normalized enhanced counter function, as distinguished from the enhanced counter function A, which is the number of secondary electrons after a given number, $N = 20$ in our case, of impacts. Normalization means that $A = e_{20}$ is divided by c_0, the number of initial "seed" electrons distributed along the contour line with a given space and phase distribution relative to the field. The data for the SEY was taken from the code [5] with a maximum equal to 1.5 at 400 eV and crossover points 50 and 1500 eV at SEY = 1, corresponding to a clean niobium

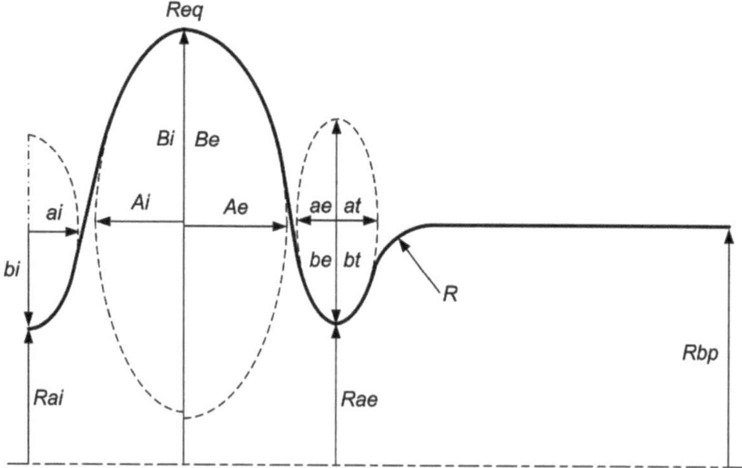

Fig. 11.1 Geometry of the iris-to-beam-pipe transition

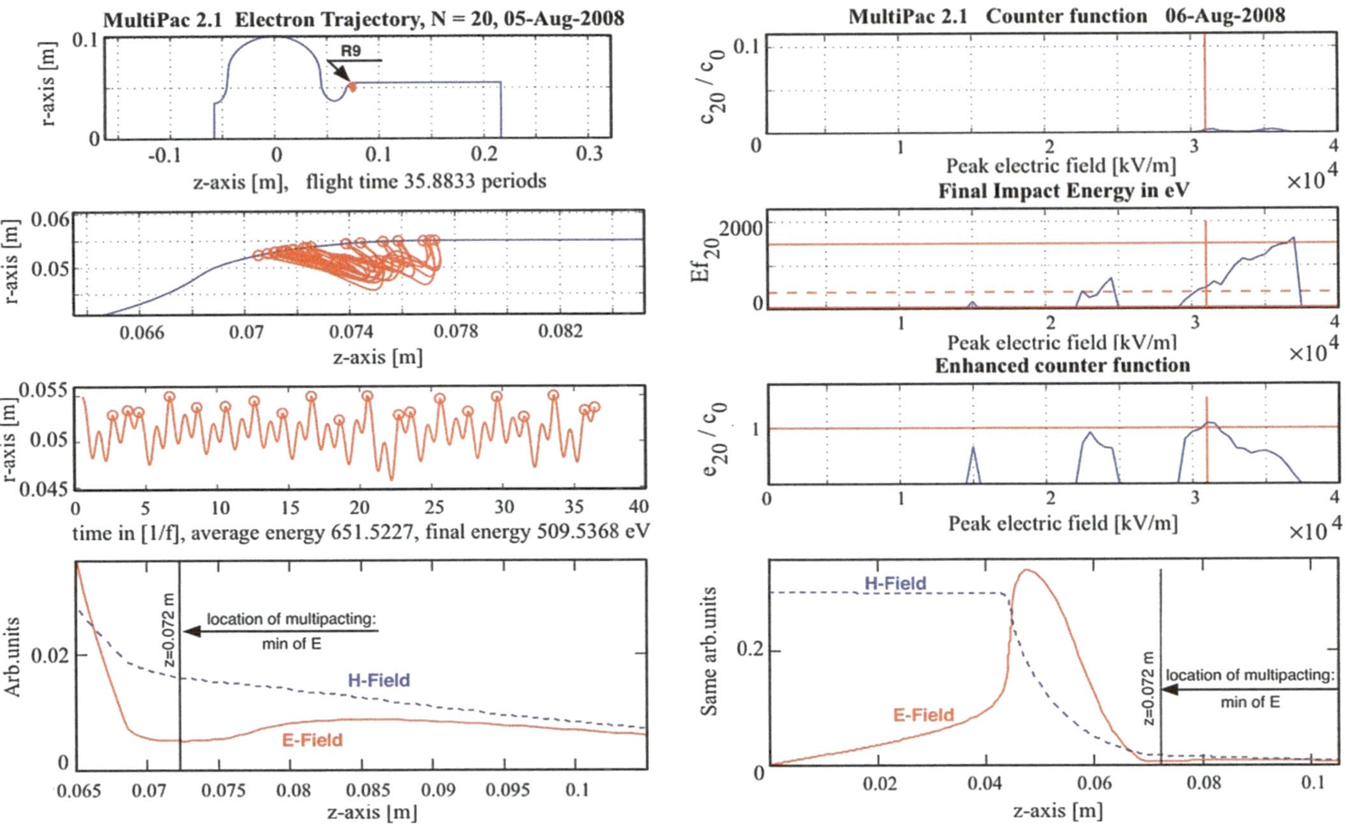

Fig. 11.2 Multipacting simulation for a cavity with an inner corner rounded with $R = 9$ mm

surface. This is a typical experimental situation in which multipactor can initially occur, but it is easily processed after some operation time with RF power when the surface becomes free of adsorbates.

Let us emphasize that the location of the multipactor between $z = 70$ and 76 mm coincides with the minimum of the electric field E shown in the lower left plot in Fig. 11.2. The whole distribution of the fields along the profile line starting from the equator ($z = 0$) is shown in the lower right plot.

An increase of the radius R to 18 mm eliminates multipacting although the very shallow minimum of E still exists (left-hand side of Fig. 11.3). This minimum disappears when the radius reaches $R = 36$ mm (right-hand side of Fig. 11.3), which is 2 times the difference between R_{bp} and R_{ae} (Fig. 11.1). This is possibly a sufficient condition to exclude multipacting in such transitions.

The variation of the radius R in the explored range changes the cavity resonant frequency by less than 1 kHz. The change of the Q factor is negligible as well.

A more detailed study indicates that there are several multipacting bands appearing at different field levels. Figure 11.4 shows the dependence of the maxima of the enhanced counter function A on the radius R and corresponding values of the peak electric field E as a function of R. Three sets of points correspond to three different bands of multipacting. Analyzed values of field levels are in the range of 25 to 35 MV/m. This range of peak electric field was chosen for illustration because it has the most distinct maximum of the function $A(R)$. Two points from Fig. 11.4 corresponding to $R = 12$ mm are further analyzed in Fig. 11.5. There are two maxima of the normalized enhanced function e_{20}/c_0 presented with corresponding impact energies and trajectories. One can see that these trajectories can be related to two kinds of multipacting: three-periodic for 25 MV/m, and two-periodic for 33.5 MV/m (with some deviations from the exact periodicity). Both kinds of multipactor are located in the flat minimum of the electric field.

11.3 Cavity with a Tapered end Port

The 9-cell Ichiro cavity [3, 6] has a tapered transition from a larger diameter cavity beam pipe to a smaller diameter end port. Figure 11.6 shows the Ichiro cavity end cell with two different transitions: one with rounded taper corners (upper half of the drawing) and the other with sharp corners (lower half). The straight beam pipe between the cell and the taper houses HOM loop couplers and an RF field probe (not shown in Fig. 11.6) and cannot be shortened. As the overall cavity length is fixed, the distance between the couplers and the flange limits the taper length to 23 mm. Rounding radii at the ends of the transition cone are equal to approximately 10.9 mm. While actual dimensions can be slightly different and the presence of the HOM couplers can somewhat change the fields in this transition, the small corrections would not affect the results in a significant way. The MultiPac simulations show the existence of resonant trajectories in the minimum of the electric field (Fig. 11.7), though again, the code doesn't calculate the normalized

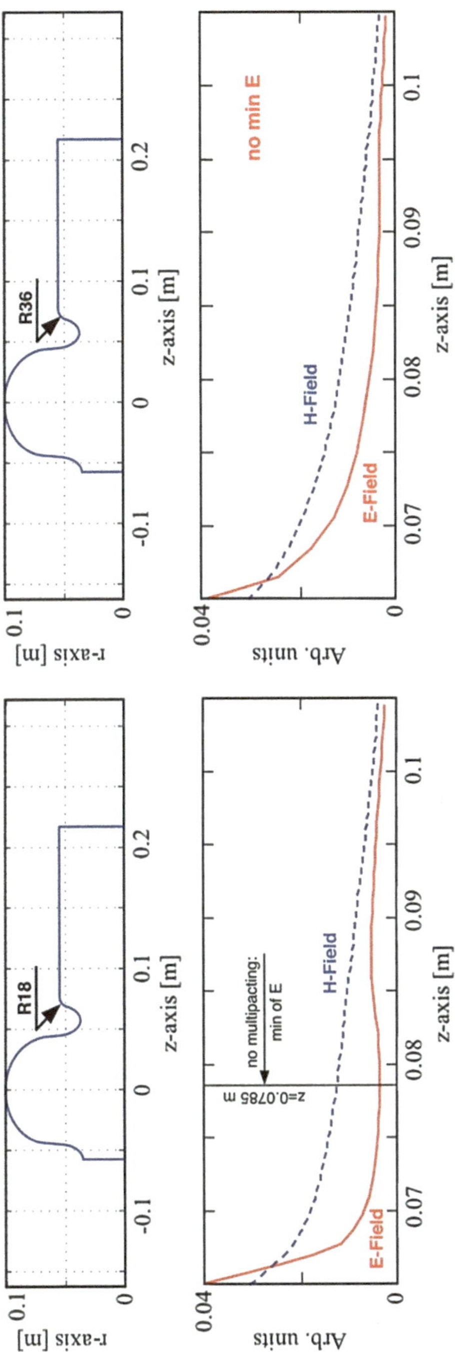

Fig. 11.3 Minimum becomes shallow at $R = 18$ mm and disappears at $R = 36$ mm. No multipactor in both cases

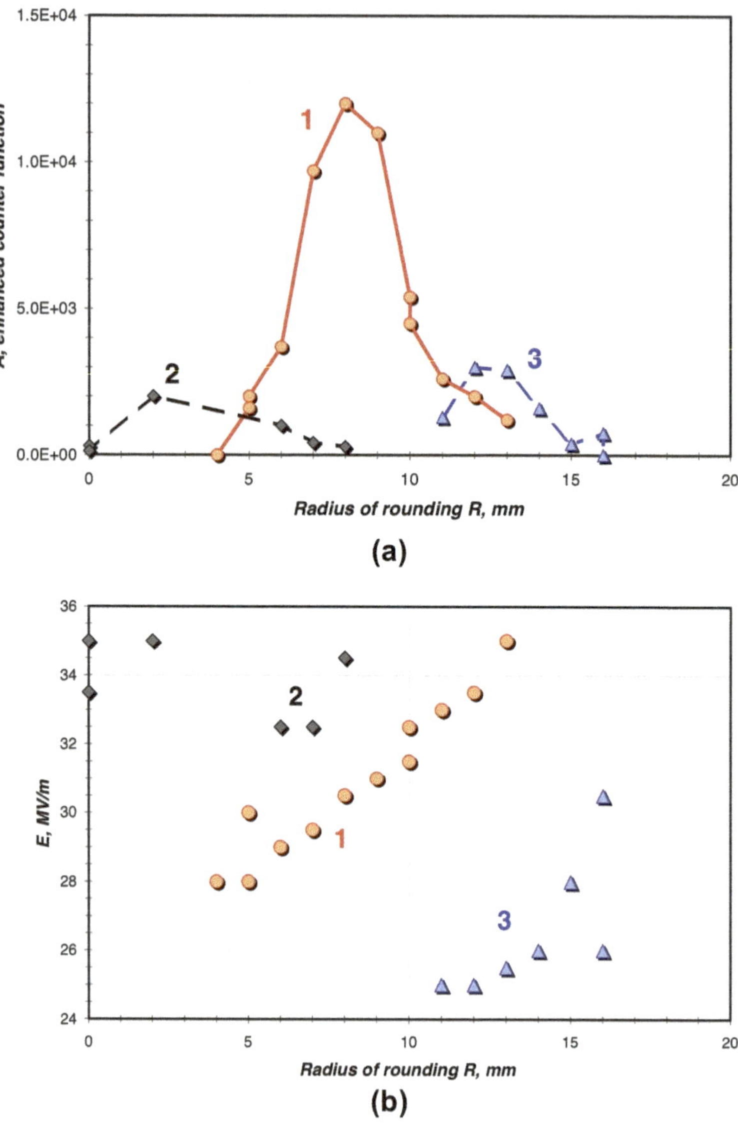

Fig. 11.4 (**a**) Dependence of maximum A on R. Three sets of points correspond to three bands of multipactor. (**b**) Dependence of E corresponding to max A on R. Three sets of points as in (**a**)

enhanced counter function of more than 1. Our results indicate that multipacting might occur at the same cavity field as reported in [3], taking into account that the field in Fig. 11.7 is the peak surface electric field, while in [3], multipactor zones are plotted as a function of the accelerating gradient.

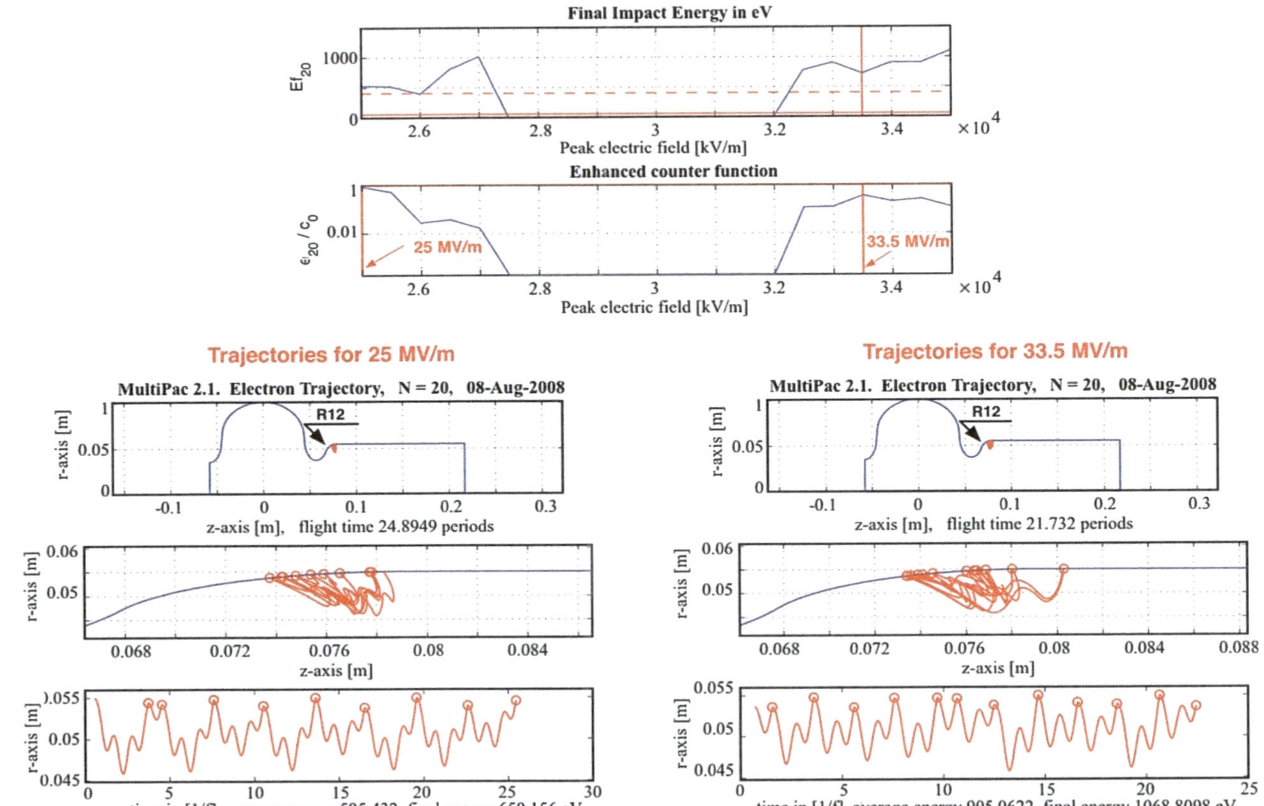

Fig. 11.5 Three- and two-periodic multipactor at two different amplitudes of the peak electric field for $R = 12$ mm

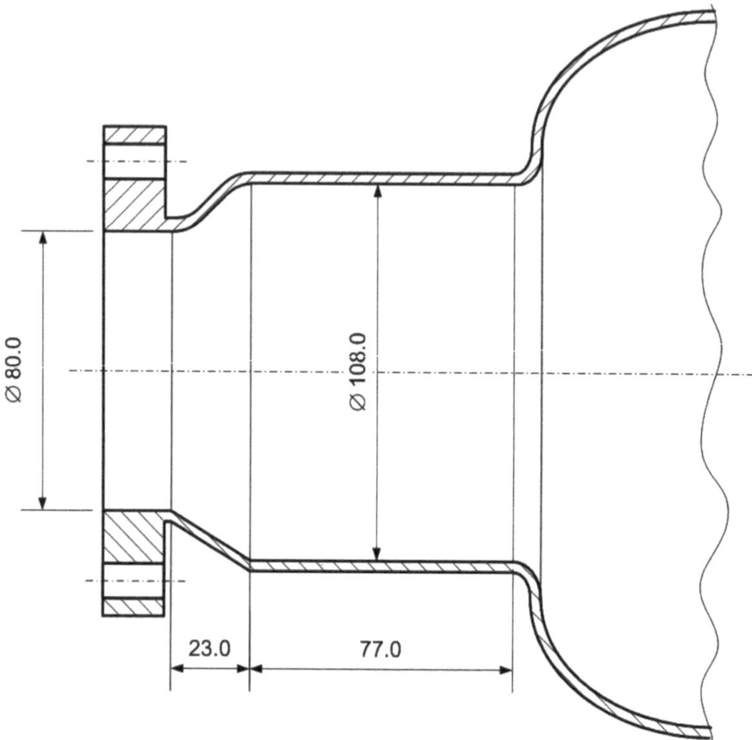

Fig. 11.6 Drawing of the Ichiro cavity end cell with the tapered port. Upper half: shape with rounded taper corners. Lower half: shape with no rounding. Dimensions in mm

Keeping the length of the taper equal to 23 mm, one can decrease the taper angle by making the rounding radii smaller, but even for the extreme case of no rounding the minimum of the electric field still exists although it becomes narrower and deeper. The only reasonable way to eliminate the minimum and thus multipacting in this geometry is to increase the length of the taper transition.

11.4 Mechanism of the Motion

Motion of electrons near the minimum RF electric field can be explained by the mechanism suggested in [7]. According to this work, acceleration of electrons in an RF field is defined by

$$\ddot{\mathbf{r}}_0 = -\nabla\Phi, \ \Phi = (\eta/2\omega)^2|\mathbf{E}|^2, \tag{11.1}$$

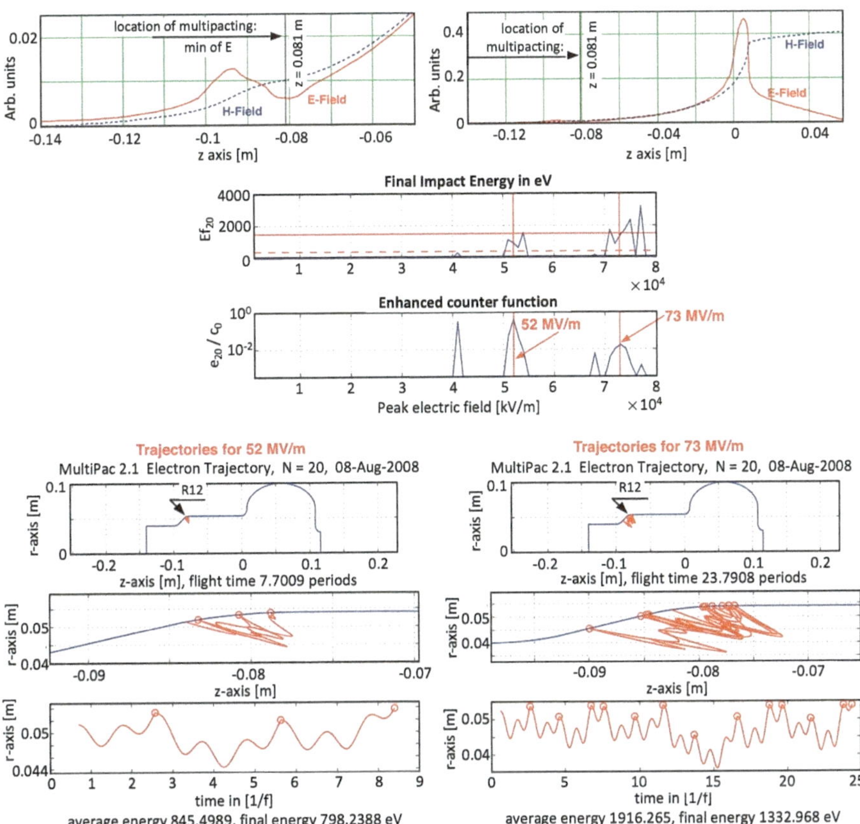

Fig. 11.7 Location of resonant trajectories in the Ichiro cavity

where $\eta = e/m$, $\mathbf{r}_0(t)$ is a slowly varying function (in terms of the RF oscillation period), as distinguished from an oscillating function $r_1(t) = A \sin \omega t$ (we omit terms connected with initial position and velocity), and Φ is the potential proportional to the square of the electric field amplitude. Function $r_1(t)$ is assumed to be much smaller than the distance over which the amplitude of the RF field changes significantly. As one can see from the above expression, electrons are pushed to the region of lower electric field amplitude (so-called Miller force). This approach was used in [8] to describe motion of electrons in a coaxial transmission line.

In the cases considered in this chapter, one has an additional effect. Wherever the electric field amplitude reaches a minimum, there is a potential well. therefore, multipacting electrons are (i) pushed toward the minimum of the electric field and (ii) can be easily trapped in that region if the potential well is deep enough. While the condition for smallness of the oscillating function may not always be valid, it does not invalidate the related physical effects but is only exactness of the formulae.

One can imagine the behavior of electrons in the cavity low electric field region as an electron wind blowing in the direction from the iris to the beam pipe. In a "calm corner" behind the iris, electrons can accumulate and multipactor can emerge if the SEY is high enough. Eliminating these "calm corners" can make the cavity free of multipacting.

11.5 Conclusion

Attraction of multipacting electrons to the minimum of electric field was explained in terms of the potential well created in the RF field [7]. We pointed out that two-point multipacting near the cavity equator (Chap. 8) also exists near the minimum (zero) of the electric field. In the case of traveling one-point multipactor (Chap. 9), electrons also move toward zero electric field at the equator. In the cases analyzed in this chapter, we see more complex trajectories of multipactor electrons in the region of minimal E.

The fact that the multipacting electrons are attracted to a minimum of electric field gives us an insight into how to avoid this phenomenon. Namely, the multipactor-free transitions between cavities and beam pipes should have shapes designed to avoid local minima of the surface electric field. In such transitions, instead of being attracted to "calm corners", electrons will be swept away in the direction of decaying field.

The principle or theorem of minimal electric field can be applied to non-elliptical RF cavities. An example of such a cavity is the so-called "balloon" variant of a single-spoke superconducting resonator proposed at TRIUMF [9]. The spoke resonator is a type of half-wave transverse electro-magnetic (TEM) mode cavity [10]. In a simple spoke resonator, one or several half-wavelength spokes are inserted into a pillbox-like shell perpendicularly to the pillbox axis of symmetry, as shown in Fig. 11.8a. Spoke resonators are sensitive to multipacting. Strong multipacting barriers were encountered during cold tests at such cavities at several laboratories. Simulations of multipactor in a simple single-spoke resonator model showed that multipacting barriers exist in a wide range of cavity voltages and that all potential multipactor locations are where the electric field reaches local minima that can be described as an RF potential well following [1]. Specifically, the first order barrier is located at the spoke bases, and higher order barriers are located at the pillbox corners. By analogy to elliptical cavities, the shell geometry was modified to be more roundish or balloon-like, as in Fig. 11.8b. In combination with re-shaping local geometry of spoke bases, this led to narrowing the multipacting barriers and pushing them to lower electromagnetic field levels. As a result, multipactor in the balloon cavity became weaker and did not present any significant problems during cold testing.

While in this book we did not explore multipacting in coaxial or waveguide transmission lines, the potential well theory [7] can be successfully applied to those cases too. As we have already mentioned, Miller force was used in [8]

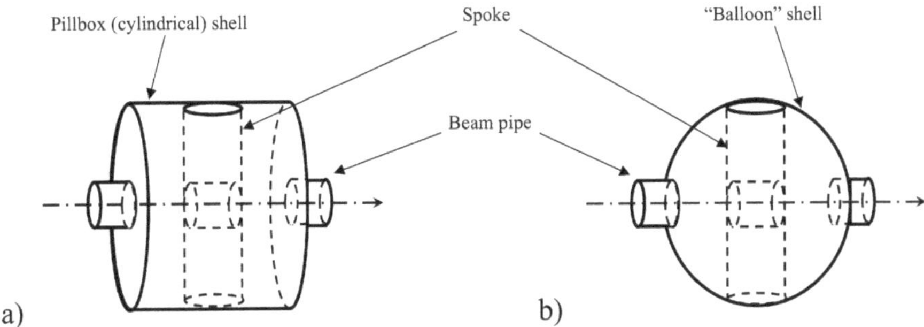

Fig. 11.8 (**a**) Pillbox and (**b**) balloon variants of a single-spoke resonator

to derive an analytic solution of electron motion in a coaxial transmission line and also was invoked in discussion of multipacting in a waveguide iris [11]. In addition, this force explains drift of multipacting electrons from the waveguide midline to the sidewalls, as observed in computer simulations [12] and, in the case of partial or full standing waves in a transmission line, migration of electrons toward the standing wave minimum. In transmission lines with a non-uniform cross section, multipacting particles will tend to move to locations where the electric field magnitude is smaller. One could say that there is an "electron wind" blowing in the direction of lower electric field along a transmission line. This electron wind can be gentle (drift is slow) if the field has a small gradient, for example near the field maximum. In this case it does not inhibit multipacting in high-field areas but only increases susceptibility thresholds there while simultaneously enhancing multipacting in lower-field regions.

References

1. S. Belomestnykh, V. Shemelin, Multipacting-free transitions between cavities and beam-pipes. Nucl. Instrum. Methods Phys. Res. A **595**, 293 (2008)
2. R.L. Geng et al., Fabrication and performance of superconducting RF cavities for the Cornell ERL injector, in *Proceedings of PAC2007, Particle Accelerator Conference*, Albuquerque (2007), p. 2340
3. Y. Morozumi, RF structure design and analysis. http://lcdev.kek.jp/ILC-AsiaWG/WG5notes/, 18 May 2007
4. P. Ylä-Oijala et al., *Multipac 2.1 – Multipacting Simulation Toolbox with 2D FEM Field Solver and MATLAB Graphical user Interface*. User's Manual, Rolf Nevanlinna Institute, Helsinki (2001)
5. P. Ylä-Oijala, D. Proch, MultiPac – Multipacting simulation package with 2D FEM field solver, in *Proceedings of the 10th Workshop on RF Superconductivity*, Tsukuba (2001), p. 105
6. T. Higo et al., Estimation of transient Lorentz detuning of ICHIRO cavity with helium jacket. ILC-Asia-2007-02. http://lcdev.kek.jp/ILCAsiaNotes/, 17 Sept 2007
7. A.V. Gaponov, M.A. Miller, Potential wells for charged particles in a high-frequency electromagnetic field. Sov. Phys. JETP **7**, 168 (1958)

8. R. Udiljak et al., Multipactor in a coaxial transmission line. I. Analytical study. Phys. Plasmas **14**(3), 033508 (2007); V.E. Semenov et al., Multipactor in a coaxial transmission line. II. Particle-in-cell simulations. Phys. Plasmas **14**(3), 033509 (2007)
9. Z. Yao, R.E. Laxdal, V. Zvyagintsev, Balloon variant of single spoke resonator, in *Proceedings of 17th Workshop on RF Superconductivity,* Whistler (2015), p. 1110
10. M. Kelly, Superconducting radio-frequency cavities for low-beta particle accelerators. Rev. Accel. Sci. Tech. **5**, 185 (2012)
11. R. Udiljak et al., Multipactor in a waveguide iris. IEEE Trans. Plasma Sci. **35**, 388 (2007)
12. E. Chojnacki, Simulation of multipactor-inhibited waveguide geometry. Phys. Rev. ST Accel. Beams **3**, 032001 (2000)

Correction to: Multipactor in Accelerating Cavities

Correction to:
V. D. Shemelin, S. A. Belomestnykh, *Multipactor in Accelerating Cavities*, Particle Acceleration and Detection, https://doi.org/10.1007/978-3-030-48198-8

The following corrections have been made to this book after the original publication:

Chapter 3:

Figure 3.1 has been replaced with the below figure with corrected references.

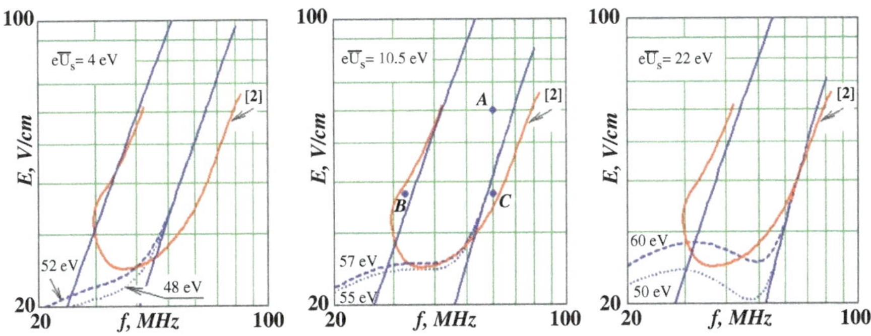

The updated online versions of the chapters can be found at https://doi.org/10.1007/978-3-030-48198-8

V. D. Shemelin, S. A. Belomestnykh, *Multipactor in Accelerating Cavities*, Particle Acceleration and Detection, https://doi.org/10.1007/978-3-030-48198-8_12

Chapter 6:

Page 55, 2nd line in section 6.3:

The text "The cell contour consists of" has been changed to "The contour of a half-cell consists of".

Chapter 7:

Page 64, 3rd paragraph:

The text "as a rule, multipactor develops in such regions" has been changed to "the case when the radial electric field is important is discussed in the next two chapters".

Chapter 8:

Page 68, The following paragraph has been added after the first paragraph:

Another effect that should be considered is the fact that electrons fall on the surface at an angle not necessarily normal to it, and SEY grows as the angle deviates from normal, see discussion in Sect. 8.7.

Page 73, 3rd paragraph

The text "further required further" has been replaced with "required further".

Page 80, 1st paragraph

The text "An interesting outcome of this analysis is the fact that the sizes of orbits of electrons participating in multipactor on the cavity equator do not depend on the size of the cavity, i.e., on frequency, and are of the order of 1 mm or less in all practical cases, as shown in Figs. 8.10 and 8.11."

has been replaced with the following text:

"An important outcome of this analysis is the fact that the sizes of orbits of electrons participating in multipactor near the cavity equator are proportional to the wave length. They are of the order of 1 mm or less for $f = 1300$ MHz, as shown in Figs. 8.10 and 8.11.

If the frequency changes, the initial normalized velocity of secondary electrons changes also: $x' = \dot{x}/\omega$, see Eqs. (8.14) and (8.15). Because the actual initial velocity does not depend on frequency, the initial normalized velocity is inversely proportional to frequency. To keep solutions of the equation (8.15) the same, the initial distance of the secondary electrons from the center of coordinates (Fig. 8.2) should increase if the frequency decreases, so that the ratio of x' to x remains the same. The actual velocity is a function of RF phase, but not explicitly of the frequency. This means that acceleration is proportional to frequency: for lower frequency actual acceleration is lower but the velocity is the same at the same RF phase because the time for acceleration is increased. Hence fields in the multipactor area are proportional to frequency similar to the acceleration. Following that we can conclude that $M = eB_0/m\omega$ does not depend on frequency. The geometrical parameter p also does not depend on frequency, because it is a ratio of the electric field components at two points near the multipactor area. So, the multipacting maps, Figs. 8.8, 8.9, 8.10, and 8.11, are valid for any frequency."

Chapter 9

Page 99, 1st line: The word "thi" has been replaced with "the".

Page 99: Figure 9.9 has been replaced with the below image with corrected labels.

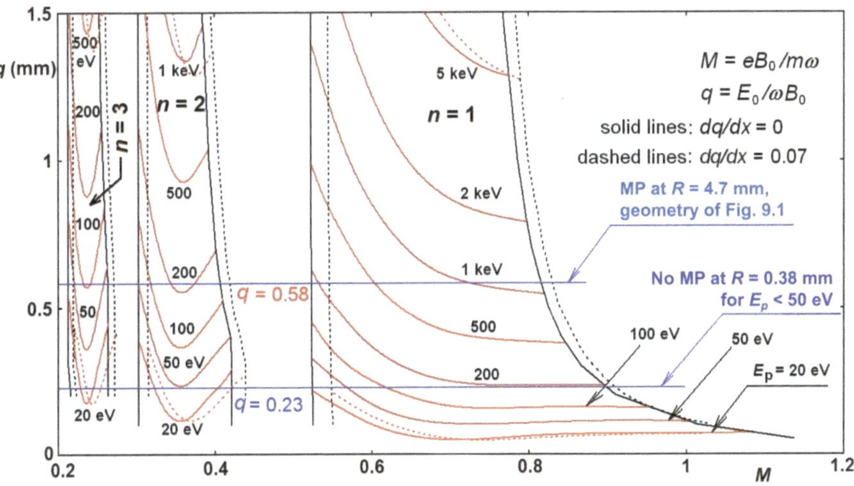

Page 101, 2nd line: The text "its velocity" has been replaced with "its starting velocity".

Chapter 10

Page 116, after equation 10.3: The text "where r is measured from the equator to the axis." has been replaced with "where r is measured from the equator toward the axis, i.e., opposite to R, z coincides with Z, as shown in Fig. 10.6".

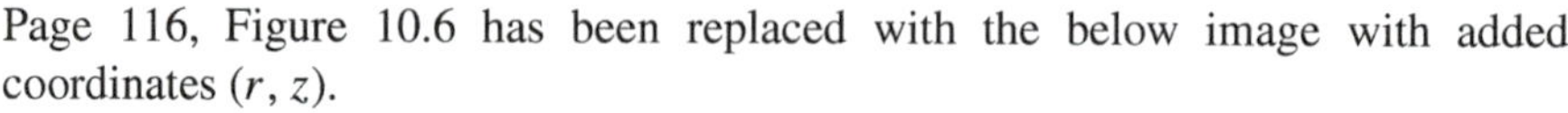

Page 116, Figure 10.6 has been replaced with the below image with added coordinates (r, z).

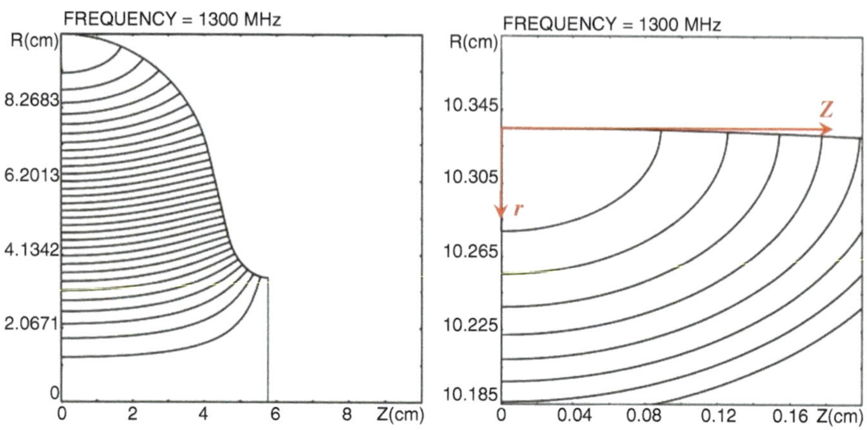

Page 117: the following sections have been added after Section 10.4

10.4.1 Multipactor in the End Cells of a Multicell RF Cavity [14]

As it was shown in [15], see also Chap. 11, multipactor electrons are attracted to the electric field minimum on the cavity surface. This minimum may shift off the middle plane if the cell is asymmetric, for example, if it has an inner cell on one side and a beam pipe on the another side (an end cell of multi-cell cavity). This shifted minimum can be treated in the same way as the minimum on the equator. The changing curvature of the surface near this point can be neglected because the distance at which the fields have to be calculated are usually much smaller than the curvature radius at this point. The only complication is the necessity to draw a perpendicular line to that point on the ellipse, while it is straightforward to draw this line from the equator.

Consider the minimum of electric field shifted to point 0 in Fig. 10.8. We need to erect a perpendicular line to this point on the ellipse and find the fields at a distance d, at point 1, and at points 2 and 3 on a straight line parallel to the tangent at point 0, at a distance d from point 1. Only the upper part of the cell is shown in Fig. 10.8. It is defined by an ellipse with half-axes A and B. We will designate coordinates of points 0, 1, ... as (Z_0, R_0), (Z_1, R_1), ...

Now we can write a system of equations to find coordinates of point 1:

$$\frac{R_1 - R_0}{Z_1 - Z_0} = \frac{A \cdot (A^2 - Z_0^2)^{0.5}}{B Z_0},$$

$$d^2 = (R_1 - R_0)^2 + (Z_1 - Z_0)^2,$$

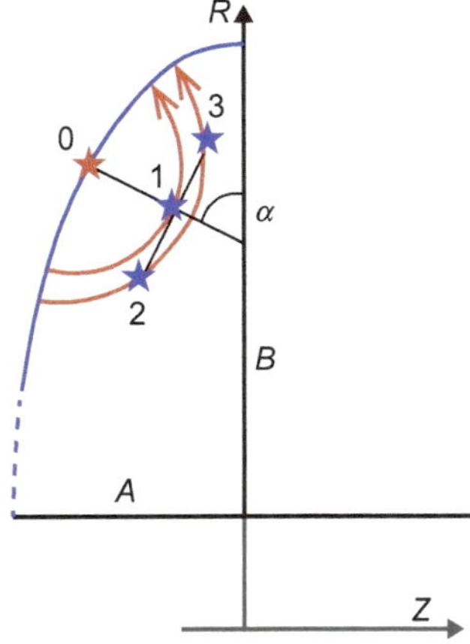

Fig. 10.8 Geometry for the case when the minimum of the electric field is shifted from the equator

and of point 2:

$$\frac{R_2 - R_1}{Z_2 - Z_1} = \frac{BZ_0}{A \cdot (A^2 - Z_0^2)^{0.5}},$$

$$d^2 = (R_2 - R_0)^2 + (Z_2 - Z_0)^2.$$

Coordinates of point 3 can be found if index 2 in the latter system changes to 3. Electromagnetic solvers usually provide E_r and E_z components of the electric field. We need to have field components parallel or normal to the tangent at point 0. Let us call them analogously to the fields in the definition of p as E'_{2r} and E'_{1z}. This transformation looks as follows:

$$E'_r = E_r \cos\alpha - E_z \sin\alpha,$$

$$E'_z = E_r \sin\alpha + E_z \cos\alpha.$$

10.4.2 Example: End Cells of the TESLA Cavity

Let us, as an example, calculate the geometrical parameter p for the end cell of the TESLA cavity. The TESLA cavity is asymmetric: it has different end cells that makes possible to extract some higher order modes otherwise trapped in the inner cells. Consider the end cell 1 with dimensions presented in [16], see Fig. 10.9 and Table 10.1. Actually, these dimensions correspond to the frequency f = 1301.369 MHz instead of 1300 MHz, as was found with SLANS [3]: the older program URMEL did not guarantee a necessary accuracy. However, this fact does not change the results substantially. If we calculate p at point 1 (Figs. 10.5 and 10.8) located on the normal to the equator, we obtain $p_2 = 0.252$ and $p_3 = 0.317$ for points 2 and

3, respectively. This is an indication that the surface electric field minimum is not at the equator. Zero of the surface electric field E_s is shifted to the left from the equator by $\Delta Z = -0.154$ mm, so that the angle $\alpha = \arctan(Z_1 - Z_0)/(R_0 - R_1) = 0.218$. However, even taking into account this small shift gives us almost equal p's, 0.285 and 0.287 for points 2 and 3, respectively. The field calculations were performed with SLANS on a 112×86 mesh.

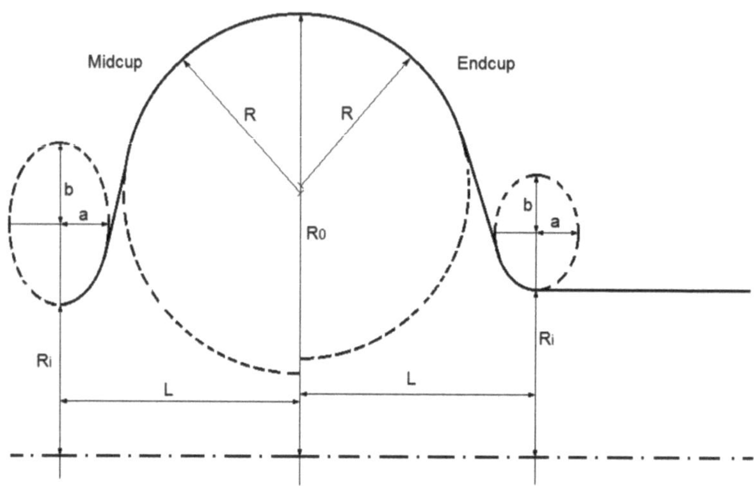

Fig. 10.9 Geometry of the TESLA end cells

Table 10.1 Dimensions (in mm) of the TESLA cavity

Parameter	Midcup	Endcup1	Endcup2
Equatorial radius R_0	103.3	103.3	103.3
External curvature radius R	42	40.34	42
Iris radius R_i	35	39	39
Horizontal half-axis a	12	10	9
Vertical half-axis b	19	13.5	12.8
Length L	57.692	56	57

Calculations for the end cell 2 give a shift $\Delta Z = -0.195$ mm and $\alpha = 0.266$ for the minimum of E_s. On the normal to the equator, we can find $p_2 = 0.240$ and $p_3 = 0.327$, but after the correction to the offset point we get $p_2 = 0.285$ and $p_3 = 0.286$. The calculation was repeated with a twice denser mesh (224×172), resulting in the shift of $\Delta Z = -0.205$ mm, and $p_2 = p_3 = 0.286$ with an accuracy of three decimal places. For the inner cells consisting of two midcaps, p = 0.286 as was found on the straight line in the plane of the equator, Fig. 10.5. It is known that multipacting in the TESLA cavities is weak and can be easily processed [17]. Note that the fields in formulas for E_r' and E_z' should be taken with the signs as calculated, but in formulas for p, shown in Fig. 10.5, absolute values of vector components should be taken.

Page 118: The following new references have been added:

14. V. Shemelin, Multipactor in the end cells of a multicell RF cavity. Phys. Rev. Accel. Beams **26**, 082001 (2023)
15. S. Belomestnykh, V. Shemelin, Multipacting-free transitions between cavities and beam-pipes. Nucl. Instrum. Methods Phys. Res. A **595**, 293 (2008)
16. D. Proch, The TESLA Cavity: Design Considerations and RF Properties, in *Proceedings of the Sixth Workshop on RF Superconductivity*, CEBAF, Newport News, Virginia, USA, 1993, p. 382
17. K. Twarowski, L. Lilje, D. Reschke, Multipacting in 9-cell TESLA cavities, *in Proceedings of the 11th Workshop on RF Superconductivity*, Lübeck/Travemünde, Germany, 2003, p. 733

Zeitfracht Medien GmbH
Ferdinand-Jühlke-Straße 7
99095 Erfurt, Deutschland
produktsicherheit@kolibri360.de